国防语言系列

航空英语综合教程

主　编　王欣然　贡卫东

副主编　丁晓松　缪　昕　张丽平　杨　静

编　者　于　阳　王华丹　刘　莹　乔　良

　　　　杨洛茜　张　力　张卉媛　郑　贞

　　　　周思平　赵宇星　高湉湉　智奕铭

南京大学出版社

前　言

谈及航空英语，人们通常想到飞行操作、航空安全、航空工程、导航通信、天气气象等专业领域。然而，不同于一般航空英语教材，本教程既是面向航空专业的专门用途英语教材，更是聚焦航空兵兵种的特色科学文化教材，既可培塑相关专业学生的岗位任职能力，也可提升读者的航空兵种文化认知。

就教材内容而言，本教程对航空兵基础知识进行系统性梳理，划分历史文化、体制编制、武器装备和使命任务四大模块，引领学生了解航空兵是什么、飞什么、做什么以及如何编制。与此同时，本教程以语言教学为根本，即在航空基本知识的框架下，以兵种专业内容为语料，突出军事英语词汇、句型和篇章的理解和应用，设计高阶性、启发性、思辨性听说与写作任务，从而帮助学生深化对航空兵种的英语认知，为未来完成航空任务打下坚实的语言基础。

本教程覆盖听、说、读、写、译五大能力，尤其注重提升学生的听说能力。为此，教程围绕各章节主题设置一系列多类型听说任务，配以丰富的影视剧、纪录片、新闻等原汁原味英文音视频资源，旨在针对性提升学生兵种英语听说能力。同时，本教程还辅以综述性和拓展性阅读文本，为学生提供航空专业背景知识。其中，综述性阅读文本一般为章节或小节主题的知识点介绍，置于每章节和小节开篇，用于预习理解，拓展性阅读文本则是章节主题的延伸，语言难度相对较高，位于每章节结尾，用于进阶提升。

教程紧扣航空兵种特色，聚焦直升机、无人机等武器装备和侦察、攻击、空中突击、空中运输等主要任务，精选阿帕奇、黑鹰、支奴干等知名直升机作为教学案例，结合其在"沙漠风暴""海神之矛""摩加迪休之战"等行动的实战运用，帮助学生直观理解航空兵种在现代战争中的核心价值和战术应用。教程末尾列出相关军事词汇，补充专业词汇与术语的中英文释义，加深专业理解。

本教程由陆军工程大学基础部大学英语课程组教员共同编撰，确保了内容的专业性和实战性。我们诚邀您加入这场语言与文化的探索之旅，让本教程成为您通向航空世界、精通军事英语的得力助手。请与我们一起，翱翔在知识的天空，开启航空英语学习的新篇章。

鉴于作者水平有限，书中可能存在错误和不足之处，敬请读者批评指正，以便我们进一步完善教材内容。

编 者

Contents

Chapter Four

What We Do: Mission and Command

Glossary

Acronyms and Abbreviations

Terms

References

Chapter One

Who We Are:
History and Culture

Alpha Overview

Since the first generation Army Aviation force has taken the arduous task, we should prove worthy of the sublime mission that the history has entrusted to us. The choice of the Army Aviation is the choice of a new start and a new challenge.

1. The Army Aviation is a branch of the Army and is an arm of modernized hi-tech force highlighting helicopters. And it is a significant power of the Army for air maneuver, fire strike and logistic support operations, acting as the link between the air and ground battle spaces in modern joint operations. Historically it used to refer to the air force when it was part of the Army before it became an independent service.

2. The main missions of the Army Aviation are to destroy enemy tanks, armored vehicles and military installations within the tactical scope, to provide aerial fire support, to transport troops and military supplies, and to conduct aerial reconnaissance and electronic warfare. The formation and development of the Army Aviation force has greatly enhanced the Army's mobility and attack firepower.

3. US Army Aviation traces its origins back to the American Civil War, where both Union and Confederate forces used hydrogen-filled balloons to direct artillery fire, marking the beginning of US military aviation and of aerial support of Army ground forces. The Army also employed balloons during the Spanish-American War and World War Ⅰ, but airplanes replaced balloons for most military purposes during the latter conflict. In the midst of World War Ⅱ, the US Army first formed an aviation unit within the artillery unit, equipped with light aircraft for reconnaissance, fire correction and communications. In the 1950s, Japan, France, the United Kingdom and Germany also established Army Aviation forces. During the Korean War, the US Army Aviation Corps began using helicopters for airlift and evacuation, as well as playing a crucial role in lifesaving and rescuing aircrews at sea. During the Vietnam War, the number of helicopters in the US Army increased to more than

ten thousand, and the organization of helicopters evolved from platoon and company to battalion and brigade, which were used to complete various combat tasks, including helicopter landing operations.

4. At the beginning of the 21st century, more than 20 countries have Army Aviation forces. The size, organization and equipment of Army Aviation varies from country to country. In October 1986, to comply with the trend of times and respond to wars under modernized conditions, the Chinese People's Liberation Army officially established the PLA Army Aviation Corps (PAAC). The PAAC is mainly equipped with attack helicopters, transport helicopters and service helicopters, and is responsible for aerial firepower assault, aircraft landing operations, aerial transportation of ground troops and materials, and battlefield service support.

5. With the continuously improved equipment performance and the increasingly diversified equipment types, the Army combat mission will be more complex, the organization will be continuously improved, and the synergy between the Army and the Air Force will be further strengthened to meet the requirements of the integrated air-to-ground operations.

Task 1　Read and answer true (T) or false (F).

1. The air force used to belong to Army. _____

2. The Army Aviation is a multipurpose tactical unit. _____

3. Airplanes had replaced balloons for most military purposes before World War Ⅰ. _____

4. More than 20 countries formed Army Aviation forces with the same equipment and organization. _____

Task 2　Fill in each of the following blanks with ONE WORD to summarize the main idea of the passage.

Para. 1 The _____ of the Army Aviation in the Army.

Para. 2 The main _____ of the Army Aviation.

Para. 3 The brief _____ of the US Army Aviation.

Para. 4 The brief introduction of the _____.

Para. 5 The _____ development of the Army Aviation.

Task 3 Read and answer the following questions.

    1. What is the role of the Army Aviation in the Army?

    2. What are the main missions of the Army Aviation?

    3. From what aspects is the PAAC introduced?

    4. What is the future development of the Army Aviation?

Task 4 Read and match the event to the war.

1. _____ American Civil War	A. the increasing number of helicopters serving in the US Army
2. _____ World War Ⅰ	B. the formation of first aviation unit within the artillery unit
3. _____ World War Ⅱ	C. the usage of helicopters for airlift and evacuation
4. _____ Korean War	D. the usage of hydrogen-filled balloons
5. _____ Vietnam War	E. the replacement of balloons with airplanes in military operations

Bravo Brief History of Aviation

While not part of the present Army Aviation Branch's heritage, United States military aviation began in 1907 with the Signal Corps Aeronautical Division and its acquisition of its first aircraft, an airplane built to Army specifications by the Wright brothers.

In 1926, the name of the air arm was changed to Army Air Corps, and then, in June 1941, the Air Corps and other Army air elements were merged to form the Army Air Forces, co-equal with the Army Ground Forces and the Army Service Forces.

During the 1930s, many Army Air Corps leaders began to experiment with strategic air operations. They advocated using air power independently of the Army ground forces to destroy enemy targets behind the lines of combat.

Following a final series of experiments with organic Army spotter aircraft conducted in 1942, the Secretary of War ordered the establishment of organic air observation for field artillery—hence the birth of modern Army Aviation—on 6 June 1942. When the United States Air Force was established as a separate service in 1947, the Army developed its light planes and rotary wing aircraft to support its ground operations. The Korean War and Vietnam War proved the growing capabilities of these aviation assets to perform a variety of missions not covered by the Air Force.

In recognition of aviation's importance, the Army established the Aviation Branch on 12 April 1983. During the 1980s, the Army introduced new, more capable helicopters, such as the UH-60 Black Hawk transport and AH-64 Apache attack helicopters, both of which saw extensive service in Operations Just Cause and Desert Storm. More recently, Army aircraft have been important assets in both Iraq and Afghanistan, transporting troops and supplies, evacuating the wounded, performing reconnaissance, and providing close air support to ground troops.

Task 1　Read and order the following events in time sequence.

☐ Organic air observation for field artillery was established, marking the birth of modern Army Aviation.

☐ The name was changed from air arm to Army Air Corps.

☐ The Aviation Branch was established in recognition of aviation's importance.

☐ The Signal Corps Aeronautical Division acquired its first aircraft built in accordance with Army specifications by the Wright brothers.

☐ The Air Corps merged to Army Air Forces with other Army air elements.

☐ Strategic air operations using air power independently were experimented and advocated.

☐ Some new helicopters with more capabilities served extensively in military operations.

🎧 1-B-1

Task 2　Listen and complete.

Army Aviation traces its origins to the American civil war when both sides used hot air balloons to send (1) _____, direct (2) _____ fire, and observe enemy (3) _____ movements. While World War Ⅰ (4) _____ aircraft had replaced balloons and the Army air (5) _____ grew from a few dozen planes to a powerful force of over 11,000 fighters and (6) _____ aircraft. During the interwar years the Army's air arm added transport and medical (7) _____, search and rescue, ground attack and (8) _____ bombing to its mission set executing these missions as the US Army air force. During World War Ⅱ, while most Army Aviation assets (9) _____ to the new US air force in 1947, the Army (10) _____ its artillery observation and reconnaissance missions providing the nucleus for a future aviation branch. Starting with the Korean War, (11) _____ aircraft increasingly became the centerpiece of Army Aviation. Coming of age during the Vietnam War, founded in 1983, today's aviation branch operates some of the most technically (12) _____ and battle tested (13) _____ in the world.

Task 3　Listen again and answer the questions.

1. What kind of missions can be conducted by the balloons?

2. What were the air arm's new missions during the interwar years?

3. What kind of missions were retained by the Army after 1947?

1-B-2

Task 4　Listen and answer true (T) or false (F).

1. The requirement of a streamlined, self-sufficient, hard-hitting forces brings new problems of control, mobility and supply. _____

2. Ground television surveillance could help the ground commander constantly understand the fluid battle situation. _____

3. Helicopters can transport units to battlefield at 35 or 40 miles an hour. _____

4. When helicopters began serving in the Army, the problems of control, mobility and supply created by the new concept of battle could be solved. _____

Task 5　Listen again and discuss the following questions in groups.

1. What are the three problems arising from maintaining a flexible Army?

2. How can aerial television surveillance help the ground commander in controlling the forces?

3. According to the text, what advantages does the helicopter offer in terms of mobility?

4. How has the introduction of helicopters helped solve the problems of control, mobility, and supply in the Army?

Task 6　Talk and present.

Introduce the history and significant role of PAAC briefly with the following sentence patterns.

> - *... is a branch of ...*
> - *... is a significant power of the Army for ...*
> - *The main missions are ...*
> - *The formation and development of the Army Aviation force has greatly enhanced ...*
> - *Army Aviation traces its origins back to ...*
> - *... also played an important role in ...*
> - *... were used to complete a variety of combat tasks including ...*
> - *the ... officially established the ...*
> - *... is responsible for ...*
> - *... were merged to form ...*

Charlie Core Competencies

Scan for listening resources

A. Provide Accurate and Timely Information Collection

Through route and area reconnaissance, movement to contact, or reconnaissance in force, Army Aviation provides the commander with accurate and timely information of enemy force. Army Aviation also provides information on ground routes, dense urban terrain and infrastructure, and man-made or natural obstacles.

B. Provide Reaction Time and Maneuver Space

Army Aviation units provide the combined arms team early and accurate warning of enemy activities, reaction time, and maneuver space to prevent surprise attacks, and the ability to rapidly develop the situation after contact with the enemy.

C. Destroy, Defeat, Disrupt, Divert, or Delay Enemy Forces

When enemy forces are in close contact with friendly ground maneuver forces, Army Aviation units attack to destroy, defeat, disrupt, divert, or delay enemy forces to enable the combined arms team to seize the initiative. When enemy forces are not in close contact with friendly forces, Army Aviation units maneuver independently from ground maneuver forces to attack enemy forces before they strike friendly forces.

D. Air Assault Ground Maneuver Forces

Air assaults are combined arms operations conducted to gain a positional advantage, envelop, or turn enemy forces that may or may not pose a threat. At the tactical level, air assault operations focus on seizing terrain, destroying enemy forces and interrupting enemy withdrawal routes.

E. Air Movement of Personnel, Equipment, and Supplies

Army Aviation units conduct air movement of personnel, leaders, critical supplies, equipment, and systems to support a variety of operations, including foreign humanitarian assistance, disaster relief, homeland defense, non-combatant evacuation routine and emergency resupply of combat units, movement of defense materials and munitions, movement of fuel, ammunition, and battlefield circulation of key leaders.

F. Evacuate Wounded or Recover Isolated Personnel

Medical evacuation (MEDEVAC) is the use of ground or air platforms by medical professionals to provide timely, efficient transfer and en-route care for the wounded, injured, or ill. Personnel recovery (PR) missions are joint operations supported by Aviation in the form of transporting PR security elements or recovery forces.

G. Enable Command and Control over Extended Ranges and Complex Terrain

Army Aviation enhances command and control by enabling the maneuver commander to better understand, visualize, describe, direct, lead, and assess operations over extended ranges and in complex terrain.

Task 1　Read and match the explanations to the corresponding competencies.

A. Provide Accurate and Timely Information Collection
B. Provide Reaction Time and Maneuver Space
C. Destroy, Defeat, Disrupt, Divert, or Delay Enemy Forces
D. Air Assault Ground Maneuver Forces
E. Air Movement of Personnel, Equipment, and Supplies
F. Evacuate Wounded or Recover Isolated Personnel
G. Enable Command and Control over Extended Ranges and Complex Terrain

1. To attack enemy forces to help the friendly ground forces seize the initiative in combat. _____

2. To provide early warning of enemy activities, and to prevent surprise attacks. _____

3. To provide enemy information and information of the combat area. _____

4. To conduct combined arms operations to gain a positional advantage, or seize terrain. _____

5. To provide transfer and en-route care for the wounded, or transport personnel recovery security elements or recovery forces. _____

6. To help maneuver commanders better command and control operations over extended ranges and in complex terrain. _____

7. To transport personnel, leaders, critical supplies, equipment, and systems

to support a variety of operations. _____

Task 2   Listen and answer true (T) or false (F).

1. Information collection helps produce intelligence and distribute information in combat._____

2. Army Aviation conducts maneuver in a wider and deeper range. _____

3. Army Aviation is not good at longer reconnaissance time on the target. _____

4. Army Aviation has a better performance in maneuver and fires compared with any other Army arms._____

5. Army Aviation improves the possibility of seizing the initiative in combat with its greater ability in information collection. _____

Task 3   Listen again and complete the text.

A successful information collection supports the production of (1) _____ and the distribution of combat information. With the advantages of increased depth and breadth of (2) _____, extended reconnaissance time over the (3) _____, greater ability to gain and maintain (4) _____, increased survivability, and enhanced maneuver, fires and (5) _____, Army Aviation could provide (6) _____ information on the enemy, terrain, local populations, (7) _____, and reaction time and maneuver space. Its competency of information collection enables the commander to confirm or deny an enemy (8) _____ and concentrate combat power at the right time to seize the initiative.

Task 4   Summarize the advantages of Army Aviation in providing combat information with the following sentence patterns. Give proper examples if possible.

- *It is good at ...*
- *It can do better in ...*
- *It enables ... with a stronger ability in ...*

1-C-2

Task 5　Listen and choose the best answer.

(　　)1. The northwestern and southwestern part of China's border line is over
_____ kilometers long.

    A. 7,600　　　　　B. 6,070　　　　　C. 6,700

(　　)2. Without the PAAC, border patrol was conducted _____ by the
border defense forces.

    A. on skis　　　　B. on foot　　　　C. by air

(　　)3. Which event marks the step of border patrol mission taken by the PLA
from ground to air?

    A. The air patrol in Xinjiang border by a PAAC force.

    B. The deployment of two helicopters to Xinjiang.

    C. The two-dimensional manner of border patrol.

(　　)4. Which of the following statement is NOT true?

    A. The PAAC has conducted 12 times of air patrols in plateau and
mountainous areas till now.

    B. Air patrols by the PAAC in border areas are to conduct security,
cover, guard and screen tasks.

    C. The PAAC has the advantages of providing early warning and
reaction time for friendly forces and conducting continuous
surveillance over enemy.

**Task 6　Listen again and discuss in groups on the significance of PAAC patrolling
border areas.**

1-C-3

Task 7　Listen and order the activities during the air landing drill.

☐ Transport helicopters flew to the predetermined airspace.

☐ The air fleet arrived at an unknown sea area after 30 minutes flying.

☐ Soldiers fell in and moved forward to the designated highland.

☐ The air fleet returned along the original course.

☐ Transport helicopters fully loaded with soldiers took off under the escort
of armed helicopters.

☐ The well-armed soldiers jumped off the helicopters and slid onto the ground.

Task 8 Listen again and complete the text.

An Army Aviation regiment of the Nanjing Military Area Command（MAC）of the PLA conducted a（1）_____ air assault drill with a troop unit in an unknown sea area，testing such dangerous and difficult subjects as（2）_____，（3）_____ on island and（4）_____ operation.

According to Zhang Wei，chief of staff of the Army Aviation regiment，at the beginning of the year，the regiment organized（5）_____ military training of dangerous and difficult subjects and effectively improved troop unit's capability. During this drill，they also focused on such subjects as（6）_____ and coordination，（7）_____ communication and contact，（8）_____ and night air landing in an unfamiliar environment.

🎧 1-C-4

Task 9 Watch and answer the questions.

1. What is the role of the transport helicopter in the joint exercise?

2. What are the benefits of rappelling from the helicopter?

3. How does the armed helicopter provide support to the special operation troops?

Task 10 Watch again and tick the competencies of PAAC shown in ZAPAD/Interaction—2021 and present your reasons.

☐ Provide Accurate and Timely Information Collection

☐ Provide Reaction Time and Maneuver Space

☐ Destroy，Defeat，Disrupt，Divert，or Delay Enemy Forces

☐ Air Assault Ground Maneuver Forces

☐ Air Movement of Personnel，Equipment，and Supplies

☐ Evacuate Wounded or Recover Isolated Personnel

☐ Enable Command and Control over Extended Ranges and Complex Terrain

Task 11 Talk and present.

Watch the video about PAAC's patrol operation and retell it focusing on the core competency. You may follow the sentence patterns listed below.

> • *With the advantages of more ... / greater / increased / enhanced ... Army Aviation could do sth. / provide sth. for sb.*
> • *Its competency of ... enables sb. to do sth.*

Delta Operational Environment

Scan for listening resources

As a member of the combined arms team, Army Aviation's ability to project power throughout all the physical domains (land, air, and maritime) are vital to joint operations.

A. Expeditionary Operation

Units are deployed into remote or harsh theaters rapidly. They must get prepared to operate with limited external resupply and sustainment for several days while conducting continuous operations.

B. Mountain Operation

With increased mobility, speed, and range, Army Aviation can effectively overcome the limitations in mountainous terrain. However, several unique weather factors may influence mountain operations, such as unpredictable wind speeds, varying wind directions, frequent updrafts and downdrafts, increased frequency of turbulence, reduced night time illumination as well as severe cold weather in winter.

C. Desert Operation

Army Aviation's speed, range, lethality and versatility secure its operational advantages in desert. But blowing dust, sand, high temperatures, reduced visibility, low contrast and extended observation ranges present challenges too.

D. Jungle Operation

In restricted terrain which is characterized by dense vegetation and jungle canopy, Army Aviation provides a significant mobility and firepower advantage over enemy ground maneuver forces. However, jungle also provides the enemy with greater concealment, and increase the vulnerability of aircraft operating at low altitudes.

E. Maritime Operations

Army Aviation provides enhanced capabilities to operate in and from the maritime domain. But rapidly changing weather conditions, sea state, low visibility, low contrast, and extended observation ranges over water also present challenges.

F. Urban Operations

Urban operations include operations in and around small built-up areas, towns, small cities, and dense urban terrain. Army Aviation can effectively overcome many of the limitations imposed on traditional ground mounted and dismounted maneuver in this complex environment. However, the challenges to successful aviation operations increase based on the scale, makeup, and complexity of each unique urban area.

G. Pandemic Zones

Army Aviation can conduct reconnaissance, convoy security, air movement of medical supplies and medical, survey or security teams, and command and control support in pandemic zones. The difficulties may include the disinfection of aircraft and the limitations imposed on aircrews flying in personal protective equipment.

H. Post-Disaster Zones

Army Aviation's capability helps overcome the lack of available road networks and other infrastructure impacted in the disaster zone. Key considerations include challenges to radio communications, limited availability of host-nation fuel support, and damage to airfields and other supporting infrastructure and limited landing zones due to debris.

I. Chemical, Biological, Radiological, Nuclear, and High-yield Explosives (CBRNE) Environments

Operations in these environments are challenging because aircrew and support personnel performance is significantly degraded when exposed to radiological, chemical or biological agents, and when operating in full protective posture.

Task 1　Read and match the risks in different environments posed upon Army Aviation forces to different operations.

A. Expeditionary Operations	B. Mountain Operations	C. Desert Operations
D. Jungle Operations	E. Maritime Operations	F. Urban Operations
G. Pandemic Zones	H. Post-Disaster Zones	I. CBRNE Environments

1. Risks of being exposed to radiological, chemical or biological agents, and limitation when operating in full protective posture. _____

2. Limited external resupply and sustainment for several days. _____

3. Unique weather factors as wind speeds, wind directions, updrafts and downdrafts, turbulence, night time illumination and severe coldness in winter, etc. _____

4. Challenges upon radio communications and fuel support, damage to airfields and other supporting infrastructure, limited landing zones. _____

5. The disinfection of aircraft and the limitations imposed on aircrews flying in personal protective equipment. _____

6. The scale, makeup, and complexity of each unique area. _____

7. Blowing dust, sand, high temperatures, reduced visibility, low contrast and extended observation ranges. _____

8. Rapidly changing weather conditions, sea state, low visibility, low contrast, and extended observation ranges. _____

9. Greater concealment for enemy and increased vulnerability of aircraft.

_____

🎧 1-D-1

Task 2   Listen and answer true (T) or false (F).

1. The exercise is held jointly by the forces from the Army and the Air Force. _____

2. The bad weather made it difficult for rotary wings to take off. _____

3. The post-landing offensive operation was an important phase of the exercise. _____

4. Blue-Side tanks were shot by some new type missiles launched by the helicopters. _____

5. The rapid air landing in the mountains enabled the Blue-Side to win precious time for victory. _____

Task 3   Listen again and answer questions.

1. What was the weather along the coast like?

2. How did the armed helicopters perform facing the challenging weather?

3. What advantage did the helicopters take in air landing in complicated mountainous terrains?

4. How did the rapid air landing benefit the Red-Side ground forces?

1-D-2

Task 4   Watch and complete the text.

Jobs don't come much (1) _____ for a helicopter in pulling hostages out of (2) _____ territory, especially when that territory is a desert. And no hostage rescue mission is more infamous than the ill-fated Desert One in 1980 to (3) _____ US citizens being held in Iran. It became helicopter hell in the desert, ending in the deaths of eight American (4) _____. One of the helicopter's greatest handicaps during the rescue operation like Desert One is the noise made by its (5) _____. The enemy can hear it (6) _____. Sand storms created by the (7) _____ called brown-outs, which can (8) _____ pilots on take-off and landing are a major cause of helicopter crashes.

Task 5   Work in groups to gather information about the *Operation Desert One* (*Operation Eagle Claw*) and discuss how a dust storm contributed to the failure.

 1-D-3

Task 6   Watch and tick the challenges caused by weather.
- ☐ fierce storm
- ☐ violent and unpredictable winds
- ☐ possibility of being too close to cliff
- ☐ losing power caused by updraft
- ☐ descent caused by downdraft
- ☐ low radar visibility

 1-D-4

Task 7   Watch and retell the challenges and solution mentioned in the video using the words and expressions below.

landing zone	heavy vegetation	bamboo
try	narrow opening	rotor blades
descend	cut	slice

Task 8 Listen and complete the text.

Army Aviation forces must be organized, trained, and equipped to meet worldwide challenges against a full range of (1) _____ . However, readiness to conduct large-scale combat operations against a peer threat is the greatest challenge today. Peer threats are (2) _____ with capabilities to oppose United States forces across multiple domains world-wide or in a specific region, where they enjoy a position of relative (3) _____ . Peer threats to aviation include (4) _____ warheads, anti-aircraft artillery, man-portable (5) _____ systems, surface-to-air (6) _____ , electronic warfare (7) _____ , cyber, chemical, biological, radiological, nuclear, and high-yield explosives (CBRNE) weapons, and manned or unmanned aircraft. They present significant challenges to unmanned aircraft, and manned aircraft above terrain flight levels.

Army Aviation must be ready to (8) _____ enemy systems to enable operational and tactical (9) _____ , and be lethal, survivable, and adaptable to provide (10) _____ to the ground commander.

Task 9 Talk and present.

Work in groups. Watch and retell the operation conducted by PACCA focusing on the operational environment, the challenges and the solutions. You may follow the sentence patterns listed below.

- *... provides the enemy with greater ...*
- *... increase the danger of ...*
- *They find ...*
- *They have to try ...*
- *They use ... to ...*

Echo Further Reading

◇ **Text A**

Birth at the Call of the Times — Let the Army "Take Off"

1. October 1 of 1984 witnessed in Beijing a world-shaking grand military review, the first National Day Parade since China's reform and opening up. Watching the awe-inspiring formations from various services and the rows of sophisticated weapons passing in front of the reviewing stand, the supreme commanders of China's armed forces could not help their joy, but meanwhile they sank deeply into another type of thought. Anyhow, they felt there was something absent in the heroic Chinese armed forces... It turned out to be the Army Aviation, the arm of force acting as trail-blazers and "low-altitude killers" in the wars like Middle East War, Vietnam War, Iran-Iraq War, Afghan War (1979) and Falklands War, which had successively taken place since 1960s.

2. On the evening of July 13, 1982, an army of 120,000 Iranian troops were storming against Basra, a city of strategic importance in south Iraq. Flood-like soldiers launched waves of overwhelming charges, and ant-like tanks were roaring ahead while spurting flows of flames. Suddenly, row upon row of soldiers fell down, and the tanks and armored fighting vehicles went "dumb" one after another. The huge waves of charges then came to a standstill, just like huge waves of water kept by a water lock.

3. It turned out to be the Iraqi armed helicopters that demonstrated their mighty power. Some helicopters fought in cooperation with the ground forces, pouring the relentless bullets, shells and rockets into the Iranian artillery positions, armored vehicles, fortifications and crowds of soldiers, just like "turrets in the air". Some helicopters, taking advantage of the terrains and surface features, approached the front by stealth — now firing hail-thick projectiles in rapid concentration, or evading the enemy gun fires in agile dispersion; now pursuing the moving enemy tanks by flight firing, or hovering in the air for the attack of armored vehicles crossing trenches and dugouts. Some helicopters hid themselves, behind blindages, and once the enemy tanks closed in, they would suddenly pull up and launch attacks on the tanks taking advantage

of their speed and altitude. And the helicopters equipped with anti-tank missiles would avoid the threats of the Iranian artillery fires and take a long-range targeting against the enemy tanks.

4. Through the smoke of modern wars, the PLA generals perceived such a reality: the emergence of the Army Aviation had not only raised the capabilities of the Army, but simultaneously integrated air with ground in combined arms operations, bringing about a structural change in the whole operational system, and hence a deep transformation of the operating patterns of the Army. Since the wide employment of armed helicopters in battle-spaces as an assault armament of the Army, the traditional combat patterns, maneuvering methods and operational theories had been confronted with new challenges.

5. Experience had proved that helicopters had a strong anti-tank capability and their loss ratio against tanks was 1:19. And it was reported that the hit rate of US AH-64 helicopters' missile attacks exceeded 90%. A dispatch of a US AH-1S armed helicopter squadron might lead to the destruction of over 130 enemy tanks while that of an AH-64 squadron might knock out over 260. And the anti-tank firepower possessed by the armed helicopter units of an Army division was capable of destroying more than 400 tanks and other armored targets, which is an equivalent to the total number equipped for a USSR (Russian) motorized infantry division. During the Syria-Israel War, the Israeli forces sent out 134 sorties of armed helicopters that launched 137 Tube-launched Optically-tracked Wire-guided (TOW) anti-tank missiles, among which 99 hit the targets.

6. Meanwhile, helicopters are characterized by strong air-maneuver capability with a maneuvering speed eight times that of motorized infantries and twenty times that of on-foot infantries. In modern wars, apart from developing the sphere of hedgehopping operations at a level lower than 100 meters, the Army Aviation also created the tactics such as "leap-frog" and "hammer-and-anvil", etc., so it was more and more valued by the militaries around the world. In the course of its comprehensive modernization, the Army should have not only tanks and armored vehicles, but also helicopters that are dominant at a "treetop level".

7. The hi-tech and three-dimensional nature of modern warfare allowed no hesitation or waiting. As the limited amount of China's armaments were completely used for self-defense, the PLA Army that had been well known for its "toughened feet" were in urgent want of its own "air cavalry". The severity of modern warfare and urgency of Chinese military's modernization called for the

birth of its Army Aviation Corps.

8. In 1985, Deng Xiao ping, Chairman of the CPC Central Military Commission, announced to the world the disarmament of one million troops. In the following year, the PLA established a new arm of force: the Army Aviation.

9. Such a disarmament and establishment marked the epoch-making step that the PLA had taken in improving its military construction level and modernized operational capabilities. From then on, the Chinese Army "sent off the battle steeds and greeted the fighting eagles".

Task 1   Read and answer true (T) or false (F).

1. The 1984 grand military review showcased the awe-inspiring formations from various arms and services, including the newly-established Army Aviation force. _____

2. Iraqi armed helicopters cooperated with the ground forces to launch fierce attack upon the enemy. _____

3. The emergence of the Army Aviation brings unprecedented challenges to the traditional combat patterns, maneuvering methods and operational theories.

_____

4. The helicopters' missile attacks of a US AH-1S armed helicopter squadron has the same hit rate as that of an AH-64 squadron. _____

5. The changed nature of modern warfare and the necessity of Chinese military's modernization called for the birth of its Army Aviation Corps.

_____

Task 2   Discuss the following questions with your partner.

1. What was absent in the heroic Chinese armed forces?

2. When was the Army Aviation established?

3. How did the Iraqi armed helicopters work to demonstrate their mighty power?

4. What reality did the PLA generals perceive through the smoke of modern wars?

5. What are the two capabilities of helicopters that have been proven by experience?

Task 3 Choose the words or phrases from the passage to summarize the main idea.

A. stealth	B. loss ratio	C. operating patterns	D. military review
E. in cooperation with	F. modern warfare	G. air-maneuver	H. maneuvering
I. battle steeds	J. structural change	K. speed and altitude	L. fighting eagles
M. armed helicopters	N. modernization	O. anti-tank capability	

Since the grand (1) _____ in 1984, the supreme commanders of China's armed forces realized that the Army Aviation, acting positively in different wars, was in urgent need.

During the Iran-Iraq War, the Iraqi (2) _____ demonstrated their mighty power, by fighting (3) _____ the ground forces, approaching the front by (4) _____, or launching sudden attacks on the tanks taking advantage of their (5) _____.

Experience had shown that helicopters had a strong (6) _____ and lower (7) _____ against tanks. They are also characterized by strong (8) _____ capability and greater (9) _____ speed. So the Army Aviation could raise the capabilities of the Army, bring about a (10) _____ in the whole operational system, and a deep transformation of the (11) _____ of the Army. The severity of (12) _____ and urgency of Chinese military's (13) _____ called for the birth of its Army Aviation Corps.

In 1985, the PLA established a new arm of force: the Army Aviation. From then on, the Chinese Army "sent off the (14) _____ and greeted the (15) _____".

Task 4 Work with your partner and summarize the history of the birth of PLA Army Aviation.

Year 1982	
Year 1984	
Year 1986	

Switching the "Battlefield" from Wastelands to Urban Areas
—Why Are Urban Operations Difficult?

During the wintertime, in a certain part of southern Liaoning province, an unidentified Army Aviation brigade of the 79th Group Army (Northern Theater Command) quietly began an airborne assault exercise. For the exercise, the force commander selected not a remote forest, but an urban area.

The Unit's commander explained the new Outline of Military Training added urban operations training content including landing and taking off from the roofs of buildings, landing in very narrow areas and electronic warfare. When conducting training according to the new Outline of Military Training and Evaluation (OMTE), the unit's Party Committee recognized that there are many buildings in urban areas and more than that they are not evenly distributed. Both the electromagnetic environment and terrain make deep assaults difficult, and serve to better temper the units' actual combat capability. During the training, a two-helicopter formation was carrying out reconnaissance. Just as Wang Xian, one of the pilots, was about to transmit information, his screens suddenly showed interference and an uneven signal, eventually losing contact with the command post. "Urban operations are 'doing something hard while doing something hard'—first, you are already carrying out reconnaissance which is itself difficult (but now are doing so in a more complicated environment)." Li Zhicheng, a battalion commander in the brigade acknowledged, "The electromagnetic environment in urban areas is complicated. It is very easy for battlefield awareness to be reduced with just a little interference—and it is very difficult to gather and transmit battlefield intelligence!"

"Aircraft XXXX, transmit your data to the command post!" After losing contact with the command post, the aircraft was ordered to act as an information transfer point to retransmit information from other units. "Because of all of the tall buildings in cities and the complicated terrain, and limited space for operations, you can only use formations of 2-3 aircraft." Brigade Training Section Chief Xu Liang explained that the OMTE significantly raised the requirements for how often equipment needed to be used in complex airspace environments, so the brigade adopted a small "networked" group with high mobility and ability to quickly reorganize to better test air communications, electronic warfare, precision attacks and other tactical capabilities.

The reporter has learned that during training the brigade alters its routes through different districts according to the terrain and changes its unit composition according to tactical requirements, cooperates in researching combined tactical issues, and strengthens training in electromagnetic spectrum management according to natural and man-made environmental factors. They also strengthened their ability to innovate tactics under the new system, and implemented courses in low-altitude navigation, search and rescue, etc.

"Find the target and begin jamming!" Following the commander's order, the "Unit" began using airborne jamming equipment to cast a wide net. Immediately, the opposing force's command network were paralyzed. At the same time, ground forces, using infrared spotting attacked the enemy positions. According to reports, this exercise generated a massive amount of database, representing an effective exploration of five operational methods to use in cities methods of conducting urban operations in all of the five tiers of Chinese cities.

Chinese Army Aviation brigades are beginning to incorporate elements of urban area training. In January 2018, China introduced a new OMTE and at the unit level, more specific instructions have been issued, emphasizing new skills that command views as necessary to improve combat capability. While China's attack helicopter forces were first developed in the 1970s to help counter tanks in an anticipated Soviet invasion from the north, with the shifting of strategic priorities — and growth of China's cities — knowledge of urban environments are now more important than the deserts and steppe terrain previously emphasized.

Helicopters are also valued for their mobility and utility as scouts and transports. However, pilots acknowledge the difficulties that come from both the more confined airspace, especially when landing or taking off, and from the complicated electromagnetic environment, such as cluttered radar returns. The unit discussed in the article, an unnamed Army Aviation brigade under the 79th Group Army, is based in Liaoning Province in northeastern China and is part of the Northern Theater Command. Notably, Liaoning is more densely populated than much of China's northeast, and units could be expected to operate in the urban sprawl connecting Dandong on the border with North Korea to the Liaodong peninsula (home to the major port city of Dalian) and curving around the coast of the Bohai Gulf toward Tianjin. The new emphasis on training in urban environments is a major step forward in increasing realism in training and matching training for what the PLA sees as likely scenarios.

"Cities have low visibility and maneuverability — the point of changing the 'battlefield' is to meet the requirements of future urban operations and raise our units' ability to carry out operations in all domains."

Task 5　Read and choose the best answer.

(　　)1. What is the primary reason the unit commander selected an urban area for the airborne assault exercise?

　　A. To comply with the new Outline of Military Training requirements.

　　B. To provide a more challenging environment for the unit.

　　C. To simulate a likely future combat scenario.

　　D. To take advantage of the unit's expertise in urban operations.

(　　)2. Which factor contributes to the difficulty of deep assaults in urban areas?

　　A. The uneven distribution of buildings.

　　B. The simple electromagnetic environment.

　　C. The flat terrain.

　　D. The lack of visibility.

(　　)3. How did the brigade address the challenges of the complicated electromagnetic environment in urban areas?

　　A. They used airborne jamming equipment to disrupt the opposing force's command network.

　　B. They altered their routes and changed their unit composition.

　　C. They conducted training in low-altitude navigation and search and rescue.

　　D. They adopted a small, highly mobile "networked" group.

(　　)4. What measures has the brigade taken during training to enhance tactical flexibility and adaptability to different environments?

　　A. Changing routes based on terrain and adjusting unit composition.

　　B. Researching combined tactical issues.

　　C. Enhancing electromagnetic spectrum management training, and innovating tactics.

　　D. All of the above.

(   )5. What is the main reason the PLA is now emphasizing training in urban environments?

    A. To counter a potential Soviet invasion from the north.

    B. To improve the combat capability of attack helicopter forces.

    C. To take advantage of the mobility and utility of helicopters.

    D. To match training for likely future scenarios.

Chapter Two

How We Are Organized:
Military Organization

Alpha Overview

Each one of us is a member of a team, a team with the singular goal of creating the most outstanding Combat Aviation Brigade in the Army.

1. Army Aviation — rotary-wing, unmanned aircraft systems (UAS) and fixed-wing — has been at the forefront of wartime and peacetime operations in the 21st century. Its inherent versatility, maneuver advantage, and warfighting effectiveness will influence all dimensions of the future battlespace. Highly motivated aviation soldiers, equipped with modern systems and trained to world class proficiency, will provide commanders at all levels an exponential increase in lethality, the leadership to harness the technological revolution of the digital battlefield, and the ability to achieve decisive victory.

2. Aviation is a component of the combined arms team, not the air component of the US Army. Its primary mission is to fight the land battle and to support ground operations. Aviation is comprised of soldiers, not airmen. It performs combat, combat support, and combat service support battlefield functions. Its battlefield leverage is achieved through a combination of reconnaissance, mobility, and firepower that is unprecedented in land warfare.

3. Army Aviation is an operations branch that provides a maneuver advantage to the Army and Joint Force commanders in unified land operations through its capabilities to overcome the constraints of limiting terrain and extended distances. And it is a significant power of the Army for air maneuver, fire strike and logistic support operations, acting as the link between the air and ground battle spaces in modern joint operations. Historically it used to refer to the air force when it was part of the Army before it became an independent service.

4. Army Aviation formations are organized, trained, and equipped to support the combined arms team at the tactical and operational levels. The Army Aviation modified table of organization and equipment (MTOE) force structure consists of four types of brigade-level aviation

maneuver organizations and two types of enabling group-level organizations. Brigade-level aviation maneuver organizations include the combat aviation brigade (CAB), the expeditionary combat aviation brigade (ECAB), the theater aviation brigade (assault) (TAB-A) and the theater aviation brigade (general support) (TAB-GS).

5. The enabling group-level aviation organizations provide the necessary support and sustainment operations to ensure the aviation maneuver brigades are capable of completing their missions. These organizations include the theater airfield operations group (TAOG) and the theater aviation sustainment maintenance group (TASMG).

6. The battalion-level and squadron-level organizations of Army Aviation consist of the air cavalry squadron (ACS), the attack battalion (AB), assault helicopter battalion (AHB), general support aviation battalion (GSAB), aviation support battalion (ASB), security and support battalion (SSB), airfield operations battalion (AOB) and theater fixed-wing (FW) battalion. Various mixes of these battalions are organized into combat aviation brigades and theater aviation brigades.

7. Although the above organizations are multi-functional, given the complexity of the operational environment and mission variables requirements, aviation brigades and battalions are usually further task-organized to meet the requirements of an assigned mission. Task organization is routinely multi-component and can include any combination of platoons, companies, and battalions organized under the aviation brigade, aviation squadron task force (ASTF), or aviation battalion task force (ABTF) headquarters. Aviation brigades and squadrons/battalions can also be task-organized with ground maneuver, joint aviation, and other non-aviation joint and Army units.

Task 1　Read and complete.

Army Aviation	
Plane type	rotary-wing, (1) _____ and fixed-wing
Personnel	soldiers, not (2) _____
Attachment	combined arms team, not (3) _____ of the US Army

（**Continued**）

Primary mission	(4) _____	
Superiority	inherent versatility, (5) _____ , and (6) _____	
Capability	combat，combat support and (7) _____	
Types of organization	4 brigade-level aviation maneuver organizations	(8) _____
		ECAB
		(9) _____
		TAB-GS
	2 enabling group-level organizations	TAOG
		(10) _____
	battalion-level and squadron-level organizations	ACS
		(11) _____
		AHB
		(12) _____
		ASB
		(13) _____
		theater fixed-wing（FW）battalion
		(14) _____
	task organization	(15) _____ ; any combination of platoons, companies, and battalions
		with aviation and non-aviation joint and Army units

Bravo Aviation Brigades and Enabling Aviation Groups

The Four Brigade-Sized Formations in Army Aviation

Combat Aviation Brigade（CAB）

The CAB is structured to synchronize operations of multiple aviation squadrons/battalions or ASTF/ABTF, ground maneuver battalions, or companies and joint aviation units. The CAB is designed to be modular and tailorable, and may be task-organized as required to support offensive, defensive, stability, or defense support of civilian authorities（DSCA）operations. The CAB includes a headquarters and headquarters company（HHC）, ACS, AB, AHB, GSAB, a Gray Eagle company equipped with 12 MQ-1C unmanned aircraft, and an ASB.

Expeditionary Combat Aviation Brigade（ECAB）

The core competencies of the ECAB are to air assault maneuver forces; position personnel, supplies, and equipment; evacuate casualties and conduct PR（personnel recovery）; and enable C2（command and control）in support of the combined arms team. The ECAB is comprised of an HHC, two AHBs, a GSAB, and an ASB.

Theater Aviation Brigade（Assault）（TAB-A）

The TAB-A augments other aviation brigades or operates autonomously at the theater level to air assault maneuver forces; position personnel, supplies, and equipment; evacuate casualties and conduct PR; and enable C2. The TAB-A consists of an HHC, four GSABs, and one ASB.

Theater Aviation Brigade（General Support）（TAB-GS）

The TAB-GS provides accurate and timely reconnaissance; positions personnel, supplies, and equipment; evacuates casualties; conducts search and rescue; and enables C2 during DSCA operations. The TAB-GS is made up of an HHC, six SSBs, and a non-standard GSAB.

The Two Enabling Aviation Groups in Army Aviation

Theater Airfield Operations Group（TAOG）

The TAOG provides airfield and air traffic services（ATS）support to the combined arms team. When deployed with AOBs, the TAOG conducts airfield management operations, provides local airspace control for the area of operations (AO) in a joint environment, and provides oversight, sustainment, and maintenance support to the theater ATS assets. The TAOG includes an HHC and up to three AOBs.

Theater Aviation Sustainment Maintenance Group (TASMG)

The TASMG is resourced to provide aviation sustainment maintenance and limited depot sustainment support at the theater level. The TASMG performs repairs and returns components/end-items to their supported units or the supply system through the National Maintenance Program. The TASMG includes a headquarters and headquarters detachment, an aviation support company (ASC), and a group support company.

Task 1　Read and give the full names of the abbreviations below.

CAB	
AHB	
GSAB	
ASB	
HHC	

 2-B-1

Task 2　Listen and choose the best answer.

Scan for listening resources

1. What is the main idea of the passage?

   A. To introduce different helicopters of the 10th Combat Aviation Brigade.

   B. To introduce the command and control of the 10th Combat Aviation Brigade.

   C. To introduce the operations of the 10th Combat Aviation Brigade.

   D. To introduce the organization of the 10th Combat Aviation Brigade.

2. Which of the following statements is **NOT** the mission of the 10th CAB?

   A. To attack the enemies and support the allies.

   B. To be ready at a moment's notice to deploy anywhere in the world.

C. To prevent a potential border crossing of migrants from Central America.

D. To conduct aviation operations that enable the 10th Mountain Division to fight and win.

3. Which of the following aircraft is **NOT** mentioned?

A. B-2 Spirit.

B. UH-60 Black Hawks.

C. CH-47 Chinooks.

D. MQ-1 Grey Eagles.

Task 3　Give the abbreviation of each unit in the title/subtitles below, then listen and complete.

The 10th Combat Aviation Brigade [(1) _____]

➢ **Five battalions**

—**1-10 Attack Reconnaissance Battalion（ARB）**

The Dragons use our Apaches to conduct (2) _____ operations in direct support of the ground forces and their commanders enabling their scheme of maneuver.

—**2-10 Assault Helicopter Battalion [(3) _____]**

The Knighthawks utilized the (4) _____ to conduct (5) _____, air movement, (6) _____ and personnel recovery, enabling ground forces to rapidly (7) _____ across the battlefield.

—**3-10 General Support Aviation Battalion [(8) _____]**

Phoenix conducts aerial mission commands, (9) _____ and (10) _____ operations using Black Hawk and Chinook helicopters enabling (11) _____ to ground maneuver forces.

—**277th Aviation Support Battalion [(12) _____]**

The Eagle's battalion stands ready to assist the 10th Combat Aviation Brigade by enabling (13) _____ through multi-class supply and distribution, 24-hour operations, signal systems networking, (14) _____ operations, health support operations and downed aircraft recovery support.

—**6th Squadrons, 6th Calvary Regiment**

The (15) _____ conduct reconnaissance, security and attack operations in support of the (16) _____ through integration of AH-64 Apaches and RQ-7 Shadows by (17) _____ teaming.

> **One Headquarters and Headquarters Company** [（18）_____]

The Renegades are trained and ready to deploy to conduct command, control, supervision, （19）_____, unit level personal service and （20）_____ of worldwide （21）_____ operations against current emerging threats providing （22）_____ to all units organically attached.

🎧 2-B-2

Task 4　Listen and answer true （T） or false （F）.

1. The 1108th TASMG is one of the four sites for the Army National Guard. _____

2. The 1108th TASMG is located in Gulfport, Mississippi and supports the upper southeast nine states including Puerto Rico and the Virgin Islands. _____

3. The 1108th TASMG has 265 full-time employees and 685 total MTOE Authorized personnel. _____

4. The 1108th TASMG supports the Army National Guard and the active duty Army. _____

5. The 1108th TASMG does major overall repair of components, painting, a lot of special projects, such as classifying parts in support of the flying customers, doing major airframe repair, etc. _____

Task 5　Listen again and complete.

Here's a quick example of how we do （1）_____ for aircraft. At the beginning, we know we have an aircraft coming in for a phase. We know there is a certain list of （2）_____ that we are going to have to either （3）_____ or （4）_____.

（5）_____ prior to the aircraft showing up, we work with the customer to get ready for the project. The customers send us all the information, including the （6）_____, the （7）_____ for the aircraft and the work order. Then, our （8）_____ analyzes it for the production control and creates the work order on our side.

When the aircraft shows up, we may have to do an initial pre-phase test flight. Then the aircraft can go into work. It might have to get washed and （9）_____. The （10）_____ has to analyze whether the components pulled off the aircraft are (11) _____ or not. We keep a (12) _____ of parts so we don't have to spend a lot of time awaiting parts when they're ordered.

Then the aircraft gets put back together and has a final (13) _____.

Task 6 Listen and figure out what the speaker thinks about multitasking and what key lesson he wants to convey. Share your understanding with your partner.

Task 7 Talk and present.

1. Introduce the organization of the 10th CAB to the whole class. You are encouraged to draw a mind map. The following expressions are helpful.

> ... comprise ...
>
> ... consist of ...
>
> ... is made up of ...
>
> ... is comprised of ...
>
> ... form part of ...

2. Suppose you were Major Andy Ratcliffe, the facility supervisor of the 1108th TASMG, the US Army National Guard. You are tasked to give an oral presentation on the general introduction of the 1108th TASMG. The general information, the organization and the capabilities can be included in your presentation. The following graph is helpful.

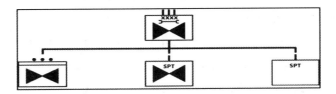

Charlie Aviation Battalions and Squadrons

As an element of the combat aviation brigade (CAB), both the air cavalry squadron (ACS) and the attack battalion (AB) provide accurate and timely information collection, provide reaction time and maneuver space, and destroy, defeat, delay, divert, or disrupt enemy forces in support of the combined arms team. The ACS typically employs its aircraft in formations as small as scout weapons teams (SWTs) of two aircraft to as large as troop or squadron formations. The AB typically employs its aircraft in formations as small as attack weapons teams (AWTs) of two aircraft to as large as company or battalion formations. The ACS is an Active Duty organization and the AB is an Active Duty and Army National Guard (ARNG) organization.

The assault helicopter battalion (AHB), a subordinate element of the CAB and expeditionary combat aviation brigade (ECAB), and the general support aviation battalion (GSAB), a subordinate element of the CAB, ECAB and theater aviation brigade (TAB), air assault maneuver forces; position personnel, supplies, and equipment; evacuate casualties; conduct personnel recovery (PR); and enable C2 in support of the combined arms team. Both the AHB and the GSAB are Active Duty, ARNG, and United States Army Reserve (USAR) organizations.

The aviation support battalion (ASB) is an element of the CAB, ECAB, and TAB. It provides aviation and ground field maintenance, network communications, resupply, and medical support. The ASB provides maintenance augmentation to aviation battalions when required. The ASB is an Active Duty, ARNG, and USAR organization.

The security and support battalion (SSB) is an element of the theater aviation brigade (general support) (TAB-GS). It is a multi-purpose aviation unit that supports a variety of federal and state missions in permissive environments by providing accurate and timely reconnaissance information; positioning personnel, supplies, and equipment; evacuating casualties; providing search and rescue; and enabling C2 in defense support of civilian authorities (DSCA) operations. The SSB is an ARNG organization.

The airfield operations battalion (AOB) provides airfield management including airfield operations, flight dispatch services, and air traffic services. The theater fixed-wing (FW) battalion provides long-range air movement for inter-theater, intra-theater, and garrison operations. The AOB is an Active Duty and ARNG organization and the theater FW battalion is an ARNG and USAR organization.

Task 1 Read and complete.

1. air _____ squadron (ACS)

2. _____ battalion (AB)

3. _____ and support battalion (SSB)

4. _____ operations battalion (AOB)

5. theater _____ battalion (theater FW battalion)

6. _____ weapons teams (SWTs)

7. Army National _____ (ARNG)

Task 2 Match the symbols with the words.

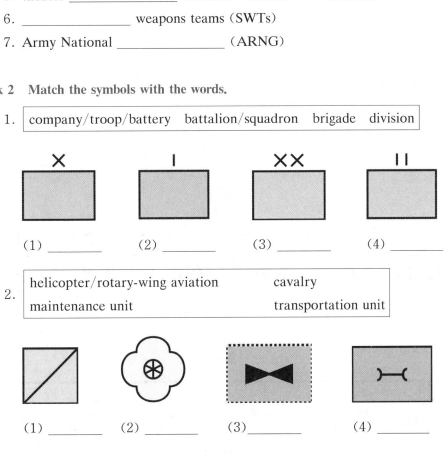

1.

| company/troop/battery | battalion/squadron | brigade | division |

(1) _____   (2) _____   (3) _____   (4) _____

2.

| helicopter/rotary-wing aviation | cavalry |
| maintenance unit | transportation unit |

(1) _____   (2) _____   (3) _____   (4) _____

Task 3　Look at the figure. Listen and complete the description of the AB organization and mission.

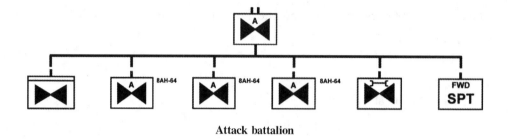

Attack battalion

The AB consists of a/(n)（1）_____ and（2）_____ company（HHC），three（3）_____ companies（ACs）equipped with eight AH-64s each，an aviation（4）_____ company（AMC），and a（5）_____ support company（FSC）.

The AB（6）_____ the following tasks: attack; zone，route，and area（7）_____; screen; guard and area（8）_____（when task-organized）; movement to（9）_____;（10）_____ in force（when task-organized）.

Task 4　Look at the figures of the ACS organization and the AHB organization. Choose one and describe it.

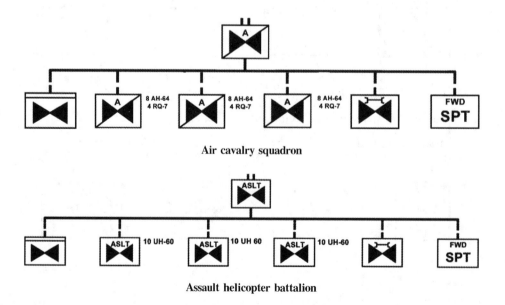

Air cavalry squadron

Assault helicopter battalion

2-C-2

Task 5　Listen and answer true（T）or false（F）.

　　1. The mission of the 602nd ASB is to provide support for the CAB. _____

　　2. The mission of the 602nd ASB includes aviation and ground field maintenance, restoration of supplies, medical support and deployment to war. _____

　　3. Prior to the modular transformation and the reorganization of US forces, the mission of the 602nd ASB was to provide logistical support to the CAB. _____

　　4. Prior to the modular transformation, the 602nd ASB had three operational sections to meet the requirements. _____

　　5. The 602nd Aviation Support Battalion SPO section was located in Korea.

_____

 2-C-3

Task 6　Listen and complete the table.

Year	Organization	Equipment	Operations
1959	The 41st (1) _____		
1988	the 1-168th (2) _____	AH-1 Cobras and OH-58 Kiowa helicopters, later (3) _____ Chinook and UH-60 (4) _____ helicopters	
2006	the 1-168th (5) _____		
2007– 2009	Aviation (6) _____- Kuwait		Operation Iraqi Freedom
2011			Operation (7) _____

Task 7　**Listen again and answer the following questions.**

1. What advance aircraft were the Battalion equipped with after the retirement of AH-1 Cobras?

2. Why did the battalion create the 1-168th GSAB?

3. Where are the units of 1-168th GSAB located?

4. What deployments have the 1-168th GSAB and its subordinate units supported?

5. In support of Operation Iraqi Freedom, what kind of units was the battalion comprised of?

Task 8　**Listen and complete.**

1st Battalion 168th Aviation Regiment (GSAB) can (1) _____ back to 1959 when it was originally (2) _____ the 41st Aviation Company in the Washington Army National Guard (3) _____ Camp Murray, Tacoma Washington. Through the next fifty years the battalion would (4) _____ several different times to meet the needs of an ever-changing national defense. In 1988 the 1-168th aviation (5) _____ an attack battalion (6) _____ AH-1 Cobras and OH-58 Kiowa helicopters. With the retirement of the AH-1 Cobra from the army inventory, the 1-168th aviation (7) _____ CH-47 Chinook and UH-60 Blackhawk helicopters. These two advance airframes would form the backbone of what would become the 1-168th Assault Battalion.

Task 9　**Talk and present.**

Suppose you are required to introduce the organization of your unit to those who just join the unit. Your introduction can be given from the following aspects.

> **History**: *be traced back to, be organized as, be designated (as), transform to, be reorganized, be activated ...*
>
> **Organization**: *consist of, be comprised of ...*
>
> **Equipment**: *be equipped with ...*
>
> **Mission**: *the mission is to, deploy to, support, in support of ...*

Delta Further Reading

The 160th Special Operations Aviation Regiment

1. The 160th Special Operations Aviation Regiment's mission is to organize, equip, train, resource and employ Army special operations aviation forces worldwide in support of contingency missions and ground force commanders. Known as Night Stalkers, these Soldiers are recognized for their proficiency in nighttime operations. They are highly trained and ready to accomplish the very toughest missions in all environments, anywhere in the world, day or night, with unparalleled precision, including airborne command and control, resupplying special operations units, search and rescue, escape and evasion activities and fire support. The unit has served in almost every U.S. conflict around the world since its formation.

History

2. Originally created as Task Force 160, the unit was formed almost exclusively from Soldiers of the 101st Airborne Division at Fort Campbell, Kentucky. In October 1981, the unit was officially designated the 160th Aviation Battalion. The regiment then became an airborne unit in October 1986 and was re-designated the 160th Special Operations Aviation Group (Airborne). The modern day 160th Special Operations Aviation Regiment (Airborne) was officially activated in June 1990. In July 2007, the regiment activated a fourth battalion to meet growing special operations forces requirements.

3. The 160th first saw combat during 1983's Operation Urgent Fury, the US invasion of Grenada.

4. In 1987 and 1988, its pilots took part in Operation Earnest Will, the protection of re-flagged Kuwaiti tankers in the Persian Gulf during the Iran-Iraq War. They flew from US Navy warships and became the first helicopter pilots to use night vision goggles and forward looking infrared devices in night combat.

5. In June 1988, the unit executed Operation Mount Hope Ⅲ. Two MH-47 crews flew 490 miles (790 km) deep into the disputed zone between Chad and Libya to retrieve a crashed Mi-24 Hind medium-attack helicopter. The following

year the Night Stalkers spearheaded Operation Just Cause in Panama, successfully executing long-range insertions to seize Torrijos and Tocumen airports and Rio Haro Airbase. They were also used in Operation Desert Storm in 1991.

6. Night Stalkers have been actively engaged in the Overseas Contingency Operations since October 2001. Today, the 160th Special Operations Aviation Regiment (Airborne) continues a sustained and active forward presence in the US Central Command area of operations at multiple locations in support of overseas contingency operations. The crews also provide support to the US Southern and Pacific commands.

7. Soldiers of the 160th pioneered the Army's nighttime flying techniques. The unit became known as the "Night Stalkers" because of its capability to strike undetected during the hours of darkness and its unprecedented combat successes. Today, Night Stalkers continue to develop and employ new technology and tactics, techniques and procedures that often migrate to other Army Aviation units, thus contributing to the overall increase in aviation capabilities. Time and again, in every major combat operation since Grenada, Soldiers of this unit demonstrate that they live by their motto, "Night Stalkers Don't Quit."

Organization

8. The 160th SOAR (A) is comprised of a regiment headquarters and five battalions. The regiment headquarters is collocated with the 1st and 2nd battalions and the Special Operations Aviation Training Battalion at Fort Campbell, Kentucky; 3rd Battalion is located at Hunter Army Airfield, Georgia; and 4th Battalion is located at Joint Base Lewis-McChord, Washington. This strategic organizational structure postures the regiment to support special operations forces mission and training requirements well into the future.

9. Each war-fighting battalion also has a strategic composition of light, medium and heavy helicopters, all highly modified and designed to meet the unit's unique mission requirements. Currently, 1st Battalion has one AH-6 Little Bird helicopter company, one MH-6 Little Bird helicopter company and two companies of MH-60 Black Hawk helicopters; 2nd Battalion has two MH-47 Chinook helicopter companies; and 3rd and 4th Battalions each have two MH-47 Chinook helicopter companies and one MH-60 Black Hawk helicopter company. The SOAR's newest aircraft include the SOF version of the Chinook heavy transport helicopter, the MH-47G, and the newest version of the Black Hawk utility helicopter, the MH-60M. Each battalion also has a Headquarters and

Headquarters Company and a maintenance company.

Assessment, Selection and Training

10. The 160th SOAR actively seeks and assesses the best-qualified aviators, crew members and support personnel in the Army. Members of this elite unit are three-time volunteers: for the Army, for airborne training and for the regiment. Upon selection, commissioned, warrant officers and enlisted Soldiers complete respective Basic Mission Qualification (BMQ) courses, known as Green Platoon, which are facilitated by the Special Operations Aviation Training Battalion.

11. A new Night Stalker arrives at a unit as a Basic Mission Qualified. After a series of skills tests, qualifications, experience and leadership, the Night Stalker is designated Fully Mission Qualified (FMQ). After three to five years as an FMQ, the Night Stalker will have the chance to assess for flight lead qualification. In a nutshell, Night stalkers are constantly training and refining their skills to ensure they are supporting the SOAR mission to the best of their ability.

12. The professionalism and capabilities of Army special operations aviation are developed through a "train as you fight" mentality. Rigorous training continues upon assignment to the line units.

13. The regiment is exceptionally resourced for extensive and realistic training conducted regularly in a variety of environments. This extensive, realistic training is the foundation of unit combat readiness.

Task 1   Read and complete.

Name change

1. In October 1981, the unit was officially designated as _____.

2. In October 1986, the unit became a(n) _____ and was re-designated as _____.

3. In June 1990,_____ was officially activated.

Combat history

1. Operation Urgent Fury (1983): the unit _____.

2. Operation Earnest Will (1987-1988): the unit's pilots became the first to use _____ and _____ in night combat.

3. Operation Mount Hope Ⅲ (June 1988): two MH-47 crews of the unit _____.

4. Operation Just Cause (1989): the unit _____.

5. Operation Desert Storm (1991): the unit was also _____.

6. Overseas Contingency Operations (since October 2001): the unit _____.

7. Today, the 160th SOAR (A) continues _____.

Organization

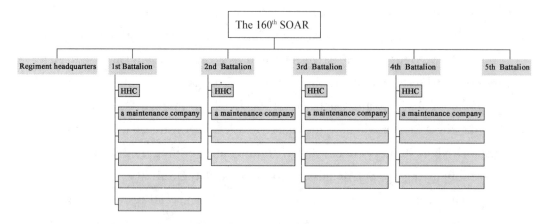

Assessment, selection and training

Task 2   Read and answer.

1. What is the mission of the 160th Special Operations Aviation Regiment?

2. Why are the soldiers of the 160th SOAR known as "Night Stalkers"?

3. What role have Night Stalkers played in various U.S. conflicts around the world?

4. What is the motto of the 160th SOAR? And what does the motto mean?

5. What is the strategic organizational structure of the 160th SOAR?

6. How does the 160th SOAR select and train its members?

7. How have Night Stalkers contributed to the development of aviation capabilities in the Army?

Task 3 Complete the sentences, and change form if necessary.

activate	pioneer	posture	retrieve
collocate	designate	spearhead	resource

1. The 75th Ranger Regiment was officially _____ as a special operations force in 1986, emphasizing their elite status within the US Army.

2. In order to complete the training exercise successfully, the logistics team needed to _____ the troops with food, water, and medical supplies.

3. The Special Forces unit was _____ and deployed to the Middle East in response to the crisis.

4. The elite Marine unit is known for _____ amphibious assault tactics that have been adopted by other military branches.

5. The search and rescue team was able to _____ the downed pilot from behind enemy lines during the mission.

6. The army decided to _____ the artillery units with the infantry for better coordination in the upcoming operation.

7. The general ordered the troops to _____ aggressively, signaling their intent to launch an offensive operation.

8. Last night, the elite marine unit _____ the amphibious assault on the enemy-held beach, securing a foothold for the main invasion force.

◇ **Text B**

The 128th Aviation Brigade

The 128th Aviation Brigade's motto "Born Under Fire" permeates everything we do as a unit in cultivating the future of Army Aviation Maintainers under the fire of rigorous training.

Our team at Fort Eustis is rich with history, experiencing birth and rebirth across the span of many decades. When Army Aviation became an established branch in 1983, Aviation logistics training continued its mission under the command of the Transportation Corps at Fort Eustis, Virginia. By 1985 the demand for unique Aviation training was clear and the US Army Aviation Logistics School (USAALS) was formed. By 1988, USAALS became a tenant activity at Fort Eustis, fully transitioning to the Aviation branch, a status it would retain until its disestablishment in 2012.

USAALS would train helicopter maintainers from initial-entry through the Advanced Noncommissioned Officer Course, the Warrant Officer Aviation Maintenance Technician Course, and the Aircraft Armament Maintenance Technician Course through four departments, the Department of Aviation Trades Training, the Department of Aviation Systems Training, the Department of Attack Helicopter Training, and the Department of Training Plans and Evaluation. By 2011, the Department of the Army activated the 128th Aviation Brigade to provide all Aviation logistics training and brigade level command and control over the Aviation community at Fort Eustis.

Currently, the 128th Aviation Brigade consists of the 1st Battalion 210th Aviation Regiment which conducts all Initial Entry Training for Apache maintainers, while also training electrical, pneudraulic, and avionics repairers for the Chinook and Black Hawk helicopters and the Warrant Officer Technician Basic and Advanced Courses. The 2nd Battalion 210th Aviation Regiment trains all Chinook and Black Hawk helicopter systems repairers along with powerplant, powertrain, and structures repairers. The 1st Battalion 222nd Aviation Regiment serves as the student battalion and manages the trainee population as they make the transition from civilian to Soldier and learn their Aviation craft. Finally, the brigade headquarters company serves to not only complete the functions of a headquarters element, but also to manage the complex operations that running a school requires. All four bodies of the 128th Aviation Brigade play a crucial role in the functioning of the unit and the production of Army maintainers and technicians.

In 2019, the 128th Aviation Brigade at Joint Base Langley-Eustis has been working hard to ensure the future success of Army Aviation as it continues to provide world-class operational support on the modern battlefield. The brigade's mission is to "generate disciplined, physically fit, technically proficient aviation maintenance soldiers and leaders" while maintaining relevancy to the operational force.

The following are some of the Brigade's maintenance training highlights.

Warrant Officer Basic and Advanced Training

The Warrant Officer Training Division (WOTD), developed and implemented more realistic, rigorous training for the Warrant Officer Basic Course (WOBC). With the addition of a new field training exercise consisting of a tactical foot march and Forward Arming and Refueling Point operation to the course, a 151A Aviation Maintenance Technician is better equipped to provide immediate support after arrival to a combat aviation brigade (CAB).

The WOTD developed a scenario that causes them to think about how Army Aviation fits into large scale combat operations against a nearpeer adversary in a multi-domain environment. Specifically, WOTD implemented a capstone training exercise, where every WOBC student plays the role of a Production Control (PC) Officer for an aviation battalion task force. That student is now responsible for developing a pre-deployment and deployment sustainment plan that enables their task force to operate in a contested and austere environment for approximately nine months. Once complete, each PC Officer then briefs their plans to the members of a Combat Aviation Brigade Maintenance Meeting, represented by Cadre, where they are put to the test to prepare them for what is to come in their first unit of assignment. The Warrant Officer Advanced Course (WOAC) will have a similar exercise, currently under development, focused on the pre-deployment and deployment sustainment plans for an entire CAB.

Enlisted Training

The 128th AB has hosted multiple Critical Task and Site Selection Boards (CTSSB) to ensure the development of quality and relevant training for the 11 Career Management Field 15 courses. The board is conducted by subject matter experts (SME) within the given CMF and relies heavily on input from various CABs. Throughout 2019, four CTSSBs were conducted for the 15D, Aircraft Powertrain Repairer, 15B, Aircraft Powerplant Repairer, 15R, AH-64 Attack Helicopter Repairer, and 15H, Aircraft Pneudraulics Repairer courses. In 2020, the brigade is scheduled to conduct CTSSBs for the following MOSs: 15N, Avionic Mechanic and 15F, Aircraft Electrician.

The 128th AB has been leading AIT trainees through a host of mentally and physically challenging situational training exercises that require all participants to successfully train and conduct 9-line medical evacuation (MEDEVAC) reports, Spot Reports, SALUTE reports, Call for Fire/Adjust Fire reports, radio calls, and visual communication techniques to standard. Once complete, the Trainees then execute a 16-event obstacle course with scenario based events throughout in a more realistic combat environment. This culminating event increases stress while also building confidence in the future Soldiers so that they develop themselves into fully mission capable Aviation Warfighters.

Task 4　Read and answer.

1. When did Army Aviation become an established branch? And when was the 128th Aviation Brigade activated?

2. What is the motto of the 128th Aviation Brigade? And what does the motto mean?

3. What is the mission of USAALS?

4. What comprises the 128th Aviation Brigade?

5. What is the mission of the 128th Aviation Brigade?

6. What is the role of WOBC students during the capstone training exercise?

7. What is special about the 16-event obstacle course?

8. What training exercise do you think is the most important to prepare trainees to accomplish their future tasks?

Chapter Three

What We Fly:
Helicopters and Drones

Alpha Overview

Helicopters are among the most adaptive and versatile weapons system worldwide. Helicopters are integral during times of war and peace.

1. Advanced military helicopters come in all shapes and sizes and are used for a wide range of purposes. Attack helicopters, for example, can reach high speeds and are well armed, while transport helicopters are bigger and can carry a much heavier payload. Multirole or utility helicopters must be good all-rounders in all environments.

2. What are the different types of advanced military helicopters and where are they used?

Attack Helicopters

3. Attack helicopters are designed for one purpose—firing on enemy troops and vehicles. Whether it is providing close air support for ground troops or destroying enemy armed vehicles and tanks, the attack helicopter is built for speed and comes heavily armed with an array of autocannons, machine guns, rockets and missiles.

4. Take the Boeing AH-64 Apache for example, which holds an M230 chain gun with 1,200 rounds, along with four hardpoints for the carrying of Hydra 70, CRV7 or APKWS air-to-ground rockets and AGM-114 Hellfire missiles. The Apache can also carry the AIM-92 Stinger air-to-air missile for aerial defence and Spike anti-tank missiles. Attack helicopters also come with advanced radars for targeting enemies and guiding projectiles.

Transport Helicopters

5. Transport helicopters are designed to fly large groups of troops, light vehicles and cargo near to the battlefield and thus need to be able to carry significant amounts of weight, both internally and underslung, as well as being relatively quick. Transport helicopters can be more useful than winged aircraft in some scenarios due to the ability for helicopters to land on most flat land without a runway.

6. The most recognizable transport helicopter is arguably the Boeing

CH-47 Chinook used by the US Army, which can travel at a max speed of around 315km/h—making it faster than many attack helicopters. However, the best transporter by weight is the Russian-developed Rostvertol Mil Mi-26. The Mil Mi-26 is a mammoth at just over 40 m in length.

Utility Helicopters

7. The utility helicopter must strike a balance to perform all kinds of activities such as conducting ground attacks, air assault, reconnaissance, troop and equipment transport, and medical evacuation. To this end, utility helicopters are highly versatile.

8. For instance, the Black Hawk has a good top speed of 295km/h, can carry a 4,100 kg external payload and can be fitted with guns, rockets, missiles and even the Volcano minefield delivery system.

Search and Rescue Helicopters

9. Above all, search and rescue helicopters must be durable, agile and packed with the latest avionics and radar systems to navigate all types of terrain and weather conditions while seeking personnel on land and at sea who need vital assistance.

10. The Agusta Westland AW101 is an iconic search and rescue helicopter in this respect. Most variants of the AW101 come with self-defence systems such as infrared jammers and laser detection.

Task 1　Read and choose the best answer.

1. What are the different types of advanced military helicopters?
   A. Transport helicopters.　　　　B. Utility helicopters.
   C. Attack helicopters.　　　　　D. All of the above.

2. Which type of helicopters does the Boeing AH-64 Apache belong to?
   A. Transport helicopters.　　　　B. Attack helicopters.
   C. Utility helicopters.　　　　　D. Search and rescue helicopters.

Task 2　Read and answer true (T) or false (F).

1. Helicopters are generally divided into utility, attack, transport and search and rescue ones. _____

2. Attack helicopters are designed to fly large groups of troops, light vehicles and cargo near to the battlefield. _____

3. Search and rescue helicopters must be durable and agile. _____

4. The Black Hawk heavily armed with an array of autocannons, machine guns, rockets and missiles. _____

5. Transport helicopters are bigger and can carry a much heavier payload. _____

6. The utility helicopter must strike a balance to perform all kinds of activities. _____

Task 3 Fill in the blanks.

1. Multirole or utility helicopters must be good _____ in all environments.

2. The Apache can carry the AIM-92 Stinger air-to-air _____ for aerial defence.

3. Transport helicopters need to be able to carry significant amounts of weight, both internally and _____.

4. Search and rescue helicopters need to seek _____ on land and at sea who need vital assistance.

Task 4 Work in groups and discuss about the following questions.

1. How many types of helicopters are there and what are they?

2. Do you know any other military helicopter? If yes, please tell which type it belongs to and give a brief introduction to it.

Bravo Brief History of Helicopters

Scan for listening resources

Helicopters are aircrafts with one or more power-driven horizontal propellers or rotors that enable them to take off and land vertically, to move in any direction, or to remain stationary in the air.

The Chinese are said to have built the granddaddy of all rotary-wing aircrafts centuries ago. Of course, this was not a real helicopter in the sense that it could carry passengers or even that it looked like any of these craft as we know them today. Rather it was a toy, which has since become known as the Chinese top.

During the mid-1500s, Italian inventor and artist Leonardo Da Vinci (1452-1519) made drawings of a flying machine that might have flapped its wings like a bird. Some experts say it inspired the modern helicopter. In 1784, French inventors named Launoy and Bienvenue demonstrated a toy to the French Academy that had a rotary-wing that could lift and fly. The toy proved the principle of helicopter flight.

The Russian-American aviation pioneer Igor Sikorsky (1889 – 1972) is considered to be the "father" of helicopters, because he invented the first successful helicopter upon which further designs were based. By 1940, Sikorsky's successful VS-300 had become the model for all modern single-rotor helicopters. He also designed and built the first military helicopter, the XR-4, which he delivered to the U.S. Army in 1941.

In 1944, U.S. inventor Stanley Hiller, Jr. (1924-2006) made the first helicopter with all-metal rotor blades that were very stiff. They allowed the helicopter to fly at speeds much faster than before. In 1946, U.S. pilot and pioneer Arthur M. Young (1905-1995) of the Bell Aircraft company designed the Bell Model 47 helicopter, the first helicopter to have a full bubble canopy and the first certified for commercial use.

Throughout history, well-known helicopter models include the UH-60 Black Hawk, the HH-60G Pave Hawk, the Sikorsky CH-53E Super Stallion, the CH-46 Sea Knight, the AH-64D Longbow Apache and the Lockheed Model 186 (XH-51).

Task 1 Read and choose the best answer.

1. What is the main characteristic of helicopters?

   A. They can only move in one direction.

   B. They can take off and land horizontally.

   C. They can only carry passengers.

   D. They can take off and land vertically.

2. Who is considered as the "father" of helicopters?

   A. Leonardo Da Vinci.   B. Igor Sikorsky.

   C. Stanley Hiller, Jr.   D. Arthur M. Young.

3. What did the toy demonstrated by Launoy and Bienvenue prove?

   A. The principle of helicopter flight.

   B. The need for further designs.

   C. The limitations of rotary-wing aircraft.

   D. The use of metal rotor blades.

4. Which helicopter model is known for having a full bubble canopy?

   A. UH-60 Black Hawk.

   B. HH-60G Pave Hawk.

   C. Sikorsky CH-53E Super Stallion.

   D. Bell Model 47.

 3-B-1

Task 2 Match the following words and phrases with the English meanings, then translate them into Chinese.

Words and phrases	English meanings	Chinese translations
1. commercial flying boat	A. the rotating armature of a motor or generator	
2. aircraft	B. a unit of power which equals to 746 watts	
3. test flight	C. the act of making something different	
4. horsepower	D. a vehicle that can fly	
5. modification	E. aircraft's trial journey	

Words and phrases	English meanings	Chinese translations
6. the Lycoming engine	F. lack of power	
7. rotor	G. a large seaplane used for commercial purposes	
8. underpowered	H. a major American manufacturer of aircraft engines	

Task 3　Listen and choose the best answer.

　　1. Which is the prototype of all rotary-wing aircraft?

　　　　A. The UH-60 Black Hawk.

　　　　B. The VS-300.

　　　　C. The Chinese top.

　　　　D. The AH-64D Longbow Apache.

　　2. Why is Igor Sikorsky called the "father" of helicopters?

　　　　A. Because he designed the Chinese top.

　　　　B. Because he made the first helicopter with all-metal rotor blades.

　　　　C. Because he designed the Bell Model 47 helicopter.

　　　　D. Because he invented the first successful helicopter.

Task 4　Listen and answer true (T) or false (F).

　　1. Igor Sikorsky made a successful series of test flights of his VS-300 in 1949-1951. _____

　　2. The VS-300 was big. _____

　　3. The VS-300 possessed the features that characterize most modern helicopters. _____

　　4. The VS-300 showed the difficulties that all subsequent helicopters would experience in the development process. _____

　　5. The VS-300 didn't lead to a long line of Sikorsky helicopters. _____

Task 5　Listen again and fill in the blanks.

　　1. He made a successful series of (1) _____ of his VS-300 in 1939-1941.

　　2. The VS-300 was small and was powered by a (2) _____ Lycoming engine.

3. Yet it possessed the features that characterize most modern helicopters: a single main (3) _____ rotor, with collective pitch, and a (4) _____ .

4. For many years, compared with (5) _____ aircraft, helicopters were (6) _____ and difficult to control, and subject to much higher dynamic stresses that caused (7) _____ and (8) _____ failure.

3-B-2

Task 6   Watch the video and answer true (T) or false (F).

1. The Stealthy Blackhawk is a top secret US helicopter. _____

2. The helicopter was first observed in 2021. _____

3. The Stealthy Blackhawk are intended to operate at night. _____

4. It seems that the stealthy helicopter has the same design as the MH-60 Blackhawk. _____

5. The helicopter is so silent that adversaries might not even react until it is too late. _____

Task 7   Watch the video again and fill in the blanks with the given words in the table.

blends into	wreckage	black paint	spot
altitudes	low noise levels	overhead	approaching
raid	shielded	tail section	

Discovery	This helicopter was first observed in 2011 during a (1) _____ in Pakistan that took down Osama Bin Laden.
Design features	• The (2) _____ of this helicopter and other design features such as reduced radar cross section and (3)_____ indicate that these special forces helicopters are intended to operate at night, when it is difficult to (4)_____ them. • These helicopters operate fast and at very low (5) _____ to avoid detection by radars. • The (6) _____ of the (7) _____ bears no resemblance to that of the MH-60 Blackhawk. • The tail rotor is (8) _____ by a disk. It looks like the noise of this helicopter easily (9) _____ any background noise.

Evidence	Neighbours of Bin Laden told that they didn't hear the helicopters (10) _____ until they were directly (11) _____.

Task 8 **Work in groups and discuss why the Stealthy Blackhawk is so difficult to spot. Try to give an answer as complete as possible.**

Task 9 **Talk and present.**

Work in pairs and make an interview. The topic of this interview can be the VS-300 helicopter, the Stealthy Blackhawk, or any other type of helicopters that you are familiar with.

> *1. The VS-300*:
> - *Test flight in* 1939–1941;
> - *Features*: *small, powered by a 65-horsepower Lycoming engine, a three-bladed rotor ...*
> - *Shortcomings*: *underpowered, difficult to control, much higher dynamic stresses*.
> *2. The Stealthy Blackhawk*:
> - *First observed in 2011 during ...*
> - *Design features*: *black paint, reduced radar cross section, low noise levels ...*

Charlie Helicopter Structure

Scan for listening resources

The structures of the helicopter are designed to give the helicopter its unique flight characteristics.

A. Airframe

The airframe（机体）, or fundamental structure, of a helicopter can be made of either metal or wood composite materials, or some combination of the two. Airframe design encompasses engineering, aerodynamics（空气动力学）, materials technology, and manufacturing methods to achieve favorable balances of performance, reliability, and cost.

B. Fuselage

The fuselage（机身）, the outer core of the airframe, is an aircraft's main body section that houses the cabin that holds the crew, passengers, and cargo. It also houses the engine, the transmission, avionics（航空电子设备）, flight controls and the powerplant（动力系统）.

C. Landing Gear or Skids

As mentioned, a helicopter's landing gear（起落架）can be simply a set of tubular metal skids（滑轮）. Many helicopters do have landing gear with wheels, some retractable（可伸缩的）.

D. Powerplant and Transmission

The two most common types of engine used in helicopters are the reciprocating engine（往复式发动机）and the turbine engine（涡轮发动机）. Reciprocating engines, also called piston engines（活塞式发动机）, are generally used in smaller helicopters. Turbine engines are more powerful and are used in a wide variety of helicopters. They produce a tremendous amount of power for their size but are generally more expensive to operate.

E. Main Rotor System

The rotor system（旋翼系统）is the rotating part of a helicopter which generates lift. The rotor consists of a mast（旋翼轴）, hub（旋翼桨毂）, and rotor blades（螺旋桨）.

F. Fully Articulated Rotor System

Fully articulated rotor blade systems provide hinges（铰链）that allow the rotors to move fore and aft, as well as up and down. When the helicopter's rotational speed changes, the rotor blades move in a certain way because of the Coriolis Effect.

G. Antitorque System

Controlled with foot pedals, the countertorque(反扭矩) of the tail rotor must be modulated as engine power levels are changed. This is done by changing the pitch of the tail rotor blades(尾旋翼叶片). This, in turn, changes the amount of countertorque, and the aircraft can be rotated about its vertical axis, allowing the pilot to control the direction the helicopter is facing.

H. Controls

The controls(控制器) of a helicopter differ slightly from those found in an aircraft. The collective, operated by the pilot with the left hand, is pulled up or pushed down to increase or decrease the angle of attack on all of the rotor blades simultaneously. This increases or decreases lift and moves the aircraft up or down. The engine throttle(油门) control is located on the hand grip at the end of the collective. The cyclic(循环控制杆) is the control "stick" located between the pilot's legs. It can be moved in any direction to tilt the plane of rotation of the rotor blades. This causes the helicopter to move in the direction that the cyclic is moved. As stated, the foot pedals control the pitch of the tail rotor blades thereby balancing main rotor torque(主旋翼扭矩).

Task 1 Read and answer true (T) or false (F).

1. The airframe can be made of either metal or composite materials. _____

2. All helicopters have landing gear with wheels, some retractable. _____

3. Turbine engines produce a tremendous amount of power for their size, but are generally more expensive to operate. _____

4. The rotor system is designed to generate lift. _____

5. Antitorque system allows the pilot to control the direction the helicopter is facing. _____

6. The collective can be moved in any direction to tilt the plane of rotation of the rotor blades. _____

Task 2　Read and choose the best answer.

(　　)1. What is the purpose of airframe design?

A. To achieve favorable balances of performance，reliability and cost.

B. To make the helicopter more visually appealing.

C. To increase fuel efficiency.

D. To reduce noise levels.

(　　)2. Which type of engine is generally used in smaller helicopters?

A. Rotating engine.　　　　　　B. Turbine engine.

C. Pistol engine.　　　　　　　D. Reciprocating engine.

(　　)3. What is the function of the fully articulated rotor system?

A. To generate lift.

B. To modulate countertorque.

C. To control the direction of the helicopter.

D. To allow rotor movement in response to rotational speed changes.

(　　)4. What do the foot pedals in a helicopter control?

A. The engine throttle control.

B. The direction that the cyclic is moved.

C. The pitch of the main rotor blades.

D. The pitch of the tail rotor blades.

 3-C-1

Task 3　Watch and answer the questions.

1. What is the definition of helicopter according to the video?

2. What is an airfoil?

3. What gets a helicopter off the ground and keeps it in the air?

4. What is a tail rotor?

 3-C-2

Task 4　Watch and choose the correct words or phrases to complete the sentences.

cockpit	cabin	tail rotors	main rotors	fuselage

1. The _____ of the helicopter are located at the bottom.

2. The _____ is the body of the helicopter.

3. The _____ is located at the rear of the helicopter.

4. The _____ is located right behind the cockpit.

5. The _____ are at the very end of the helicopter.

 3-C-3

Task 5   Watch and match the terms with their description.

1. Collective	a. controls the turning of the helicopter to the left or right by adjusting the tail rotors
2. Cyclic	b. moves the helicopter up or down by adjusting the swashplate
3. Tail rotor pedals	c. moves the helicopter forward, backward, left or right

 3-C-4

Task 6   Watch and answer the questions.

1. Why is Apache called a flying tank?

2. What are the duties of the Apache pilot and co-pilot respectively?

3. Why do both sections of the cockpit include flight and firing controls?

4. How do the controls manipulate the rotors?

5. On Longbow Apache, what equipment are included to keep the pilots informed?

Task 7   Watch again and fill in the blanks.

The Apache helicopter is a revolutionary development in the history of war. It's essentially a flying tank, a helicopter designed to survive heavy attack and (1) _____ massive damage. It can zero in on specific targets day or night, even in terrible weather. Apache Controls. The Apache (2) _____ is divided into two sections, one directly behind the other. The pilot sit in the rear section and the co-pilot gunner sits in the front section. As you might expect, the pilot (3) _____ the helicopter and the gunner aims and fires the weapons. Both sections of the cockpit include flight and firing (4) _____, in case one pilot needs to take over full (5) _____. The pilot flies the Apache using collective and (6) _____ controls，similar to ones you would find in any other helicopter. The controls manipulate the（7）_____ using both a mechanical hydraulic system and a digital stabilization system. The digital stabilization system fine-tunes the powerful hydraulic system to keep the helicopter flying smoothly. The stabilization system can also keep the helicopter in an automatic (8) _____ position for short periods of time. On the longbow Apache，three display (9) _____ provide the pilot with most（10）_____ and flight information. These digital displays are much easier to read than traditional instrument dials. The pilot simply presses buttons on the side of the display to find the information he or she needs.

Task 8　Talk and present.

Work in pairs and introduce the Apache controls to your partner. You may follow the instructions：

- *describe the roles of the pilot and co-pilot，emphasizing their respective responsibilities*
- *discuss the flight and firing controls in both sections of the cockpit*
- *explain the collective and cyclic controls used for flying，detailing the mechanical hydraulic and digital stabilization systems*
- *discuss the display panels that provide navigational and flight information*

Delta Flight Preparations

Scan for listening resources

This chapter explained the importance of preflight and safety when conducting helicopter ground operations.

Preflight

Before any flight，ensure the helicopter is airworthy（适航的）by inspecting it. Remember that it is the responsibility of the pilot in command （PC）to ensure the aircraft is in an airworthy condition. In preparation for flight，the use of a checklist is important so that no item is overlooked.

Engine Start and Rotor Engagement

During the engine start，rotor engagement，and systems ground check, use the manufacturer's checklists. If a problem arises，have it checked before continuing. Prior to performing these tasks，however，make sure the area around and above the helicopter is clear of personnel and equipment. Position the rotor blades so that they are not aligned with the fuselage. This may prevent the engine from being started with the blades still fastened. For a two-bladed rotor system，position the blades so that they are perpendicular（垂直的）to the fuselage and easily seen from the cockpit（驾驶室）. Helicopters are safe and efficient flying machines as long as they are operated within the parameters（范围）established by the manufacturer.

Safety in and Around Helicopters

People have been injured，some fatally，in helicopter accidents that would not have occurred had they been informed of the proper method of boarding or deplaning（下飞机）. A properly briefed passenger should never be endangered by a spinning rotor. The simplest method of avoiding accidents of this sort is to stop the rotors before passengers are boarded or allowed to depart. Because this action is not always practicable，and to realize the vast and unique capabilities of the helicopter，it is often necessary to take on passengers or have them exit the helicopter while the engine and rotors are turning. To avoid accidents，it is essential that all persons associated with helicopter operations，including passengers，be made aware of all possible hazards and instructed how those hazards can be avoided.

Task 1   Read and answer true (T) or false (F).

1. Before any flight, the aircrew should evaluate the helicopter by making a checklist. _____

2. Prior to engine start, aircrew should make sure the area around and above the helicopter is clear of personnel and equipment. _____

3. The simplest method of avoiding accidents caused by rotors is to send the rotors spinning after passengers are boarded or allowed to depart. _____

Task 2   Read and choose the best answer.

(   )1. Who should inspect the helicopter to ensure it is airworthy before any flight?

    A. The chief mechanic.

    B. The ground crews.

    C. The pilot in command (PC).

    D. The co-pilot.

(   )2. What should be done before engine start?

    A. Ensure the area around and above the helicopter is clear of personnel and equipment.

    B. Align the rotor blades with the fuselage.

    C. Use the manufacturer's checklists.

    D. Turn off all the systems in the helicopter.

(   )3. How should the rotor blades be positioned for a two-bladed rotor system during ground operations?

    A. Perpendicular to the fuselage.

    B. Aligned with the fuselage.

    C. Parallel to the fuselage.

    D. In a random position.

(   )4. What is the simplest method of avoiding accidents involving spinning rotors when passengers are boarding or departing?

    A. Stop the rotors before passengers are allowed to board or depart.

    B. Instruct the passengers to climb over the rotor blades.

    C. Position the passengers far away from the helicopter.

    D. Ignore the hazards associated with helicopter operations.

3-D-1

Task 3　Watch and tick the items in the pilot's kit.

☐ helmet	☐ rifle	☐ carbine	☐ body armor	☐ thermal jacket
☐ ration	☐ heliograph	☐ harness strap	☐ radio	☐ flare
☐ sidearm	☐ mine tape	☐ medical kit	☐ strops	☐ beret

 3-D-2

Task 4　Listen and complete the task description.

Preflight Inspection on AH-64

Crew actions. 1. The (1) _____ (PC) is responsible for ensuring that a preflight inspection is conducted. The PC may direct the pilot to complete (2) _____ of the aircraft preflight inspection as applicable, and will verify that all checks have been completed. The PC will report any aircraft (3) _____ that may affect the mission and will ensure that the appropriate information is entered on forms. 2. The PC will ensure a (4) _____ inspection is complete prior to flight. 3. The pilot will complete the (5) _____ elements and report the results to the PC.

Procedures. 1. Consider the helicopter armed and approach it from the side to avoid danger areas. Ensure that the aircraft is in an (6) _____ safe status and follow grounding procedures prior to continuing further with the preflight. 2. Refer to (7) _____ throughout the conduct of the aircraft preflight inspection. Comply with the preflight checks contained in the checklist and standing operating procedure (SOP) as applicable. 3. As applicable, the PC will ensure that all pertinent load maintenance panel (LMP), (8) _____ (COMSEC), and global positioning system (GPS) key data has been loaded into the aircraft.

3-D-3

Task 5　Watch the preflight briefing and complete the sentences.

　　1. Do not put anything under your seat. It is made to _____.

　　2. If you have any backpacks or personal items, make sure that _____ in front of you.

　　3. You must _____ at all times during this flight.

　　4. In the case of emergency, if you need to get out of the aircraft, the _____ is your main emergency exit.

　　5. To open the cabin door, you grab the handle and _____.

　　6. Hearing protection is always required during flight and make sure that if it's cold weather, you always _____.

　　7. You have three first aid kits, _____, one survival kit and one crash X.

Task 6　Talk and present.

　　Brief your partner about helicopter safety focusing on the following aspects:

- *the proper use of seatbelts, doors, and headsets / intercom system*
- *the safe entry and exit paths*
- *placement of all passenger personal items*
- *the location of the fire extinguisher, survival equipment*

Echo Attack Helicopters

Scan for listening resources

An attack helicopter is an armed helicopter with the capability of engaging targets on the ground, such as enemy infantry and armoured fighting vehicles. Due to their heavy armament, they are sometimes called helicopter gunships.

Weapons used on attack helicopters can include autocannons, machine guns, rockets, and guided missiles such as the Hellfire. Many attack helicopters are also capable of carrying air to air missiles, though mostly for purposes of self-defense. Today's attack helicopter has two main roles: first, to provide direct and accurate close air support for ground troops, and the second, in the anti-tank role to destroy enemy armor concentrations. Attack helicopters are also used to supplement lighter helicopters in the armed scout role.

Attack helicopters come in various sizes, from the small and maneuverable AH-6 to the heavily armed advanced models like the AH-64 Apache and Ka-52 Alligator. The Boeing AH-64 Apache is a four-blade, twin-engine attack helicopter with a tailwheel-type landing gear arrangement, and a tandem cockpit for a two-man crew. The pilot in command usually occupies the rear seat while the front seat is the copilot/gunner position; however, all weapons systems can be fired from either position.

The AH-64 is designed to endure front-line environments and to operate during the day or night and in adverse weather via its avionics and onboard sensor suites. These systems include the Target Acquisition and Designation System, Pilot Night Vision System (TADS/PNVS), passive infrared countermeasures, GPS, and Integrated Helmet and Display Sight System (IHADSS). It is one of the world's most advanced battlefield attack helicopters. It's a tank killer and one of the fastest and stealthiest attack helicopters in the world.

Task 1   Read and choose the best answer.

(   )1. What is the primary role of an attack helicopter?

        A. Engaging targets in the air.

        B. Providing air support for ground troops.

        C. Transporting troops to the battlefield.

        D. Conducting reconnaissance missions.

(   )2. What types of weapons can be used on attack helicopters?

        A. Rifles, pistols, and hand grenades.

        B. RPGs, mortars, and land mines.

        C. Flamethrowers, tear gas, and air-to-air missiles.

        D. Autocannons, machine guns, and guided missiles.

(   )3. Which attack helicopter is known for its maneuverability and small size?

        A. AH-6.   B. AH-64 Apache.

        C. Ka-52 Alligator.   D. AH-1.

(   )4. What is the role of the copilot/gunner in the AH-64 Apache?

        A. Flying the helicopter.

        B. Navigating the helicopter.

        C. Monitoring the avionics and sensors.

        D. Operating the weapons systems.

(   )5. Which systems enable the AH-64 Apache to operate in adverse weather conditions?

        A. Target Acquisition and Designation System and Pilot Night Vision System.

        B. Passive infrared countermeasures and GPS.

        C. Avionics and onboard sensor suites.

        D. IHADSS and tailwheel-type landing gear arrangement.

🎧 3-E-1

Task 2   Watch and answer true(T) or false(F).

1. Boeing delivered the first AH-64 Apache to the US Army in 1984. _____

2. The US Army has received over 2,200 Apache attack helicopters. _____

3. AH-64 completed 1.3 million fight hours in special operations. _____

4. The Apache attack helicopters served in combat during conflicts in Panama, Iraq, Turkey, Bosnia, Kosovo, etc. _____

Task 3　Watch again and complete the specifications.

AH-64 Apache attack helicopter	
Rate of Climb	(1) _____ m / min
Maximum Speed	(2) _____ km / h
Cruise Speed	(3) _____ km / h
Ferry Range	(4) _____ km
Service Ceiling	(5) _____ meters
Endurance	(6) _____ hours (7) _____ minutes
Weight	(8) _____ kg
Maximum Takeoff Weight	(9) _____ kg

🎧 3-E-2

Task 4　Tick the armament that Apache is equipped with.

☐ Hellfire missiles

☐ Mark 46 torpedoes

☐ Longbow radar

☐ FFAR rocket pods

☐ Hydra rocket pods

☐ machine guns

☐ miniguns

Task 5　Watch the video and complete the text.

The US Army's Apache helicopter has three primary weapons systems: Hellfire missiles that can pierce (1) _____ two inches thick from five miles away. Hydra rocket pods that each fire 19 (2) _____ and when pilots have to take on targets that are closer, a Bushmaster machine gun, "This is the Bushmaster 30-millimeter chain gun, big (3) _____." This weapon is (4) _____ up to nearly a mile away. The machine gun can unleash 1,200 (5) _____. Its (6) _____ is over 600 rounds in the minute. Technicians (7) _____ the gun on the Apache's underbelly, then they connect it to infrared sensors and special cameras that feed information to the pilots.

Wherever the pilot looks, the machine gun (8) _____ points.

"If I want to shoot something 90 degrees off my right, quite literally all I have to is turn my head, put that cross hairs on that target, squeeze the trigger and put rounds on that target."

Task 6　Watch again and answer the questions.

1. How many square miles can the Longbow radar scan in seconds?
2. What is the weight of the Longbow?
3. How can the Longbow scan for objects over the battlefield?
4. How can the Longbow identify whether the object is friendly or not?
5. Once confirmed as hostile, what can the Apache do to the target?

Task 7　Talk and present.

Work in pairs. Introduce WZ-10 with the information provided in the table.

直-10 直升机	
制造商	昌河飞机工业公司
加入现役	2010 年
性能参数	空重:5 540 千克 爬升率:12 米/秒 最大速度:300 千米/小时 巡航速度:270 千米/小时 转场航程:800 千米 实用升限:6 400 米 最大起飞重量:6 000 千克
驾驶舱	采用串列双座设计,前座为驾驶席,后座为副驾驶(射击员),两舱各自独立,使其遭敌火命中时,不容易同时波及两个座舱
武器系统	57 毫米/90 毫米的火箭吊舱,130 毫米的火箭弹,以及架设在机鼻下方的机炮(航炮)
功能	为友军提供近距支援,反装甲和对抗敌机

Foxtrot Utility Helicopters

Scan for listening resources

A utility helicopter is a multi-purpose helicopter. A utility military helicopter can be utilized for ground attack, air assault, cargo, MEDEVAC, command and control, and troop transport. The most iconic and notorious utility helicopter is the UH-60 Black Hawk. It is a four-bladed, twin-engine, medium-lift utility helicopter manufactured by Sikorsky Aircraft. It was designed to replace the Bell UH-1 Iroquois as the Army's tactical transport helicopter. The UH-1 Iroquois served admirably but was proved too vulnerable when landing in combat during the Vietnam War. More than 7,000 total UH-1s served in Vietnam, but more than 3,300 of them were ultimately shot down or destroyed, killing 2,100 pilots and crew members. The UH-60 was purposely built to do one thing—survive. The UH-60 entered service with the US Army's 101st Combat Aviation Brigade of the 101st Airborne Division in June 1979. This was followed by the fielding of electronic warfare and special operations variants of the Black Hawk. Improved UH-60L and UH-60M utility variants have also been developed. The following are the specifications of the UH-60L.

Crew	2 pilots (flight crew) with 2 crew chiefs/gunners
Maximum Speed	295 km/h
Cruise Speed	278 km/h
Ferry Range	2,220 km
Service Ceiling	5,790 m
Rate of Climb	4.5 m/s
Weight	4,819 kg
Maximum takeoff weight	10,660 kg
Payload	Internal load 1,200 kg, external on sling 4,100 kg, up to 11 fully-equipped troops
Armament	Guns: 2×7.61 mm (0.30 in) M24OH machine guns or

	2×7.62 mm $(0.30$ in$)$ M134 minigun or
	$2 \times .50$ in $(12.7$ mm$)$ GAU-19 gatling guns
	Rockets: 70 mm $(2.75$ in$)$ Hydra 70 rockets
	Missiles: AGM-114 Hellfire laser guided missiles,
	AIM-92 Stinger air-to-air missiles

Task 1　Answer true(T) or false(F) based on the above passage.

1. A utility military helicopter is primarily used for air assault. _____

2. The UH-60 Black Hawk was specifically built to survive in combat situations. _____

3. The UH-60 Black Hawk has only one variant, the UH-60L. _____

4. The UH-60 Black Hawk has a maximum takeoff weight of 4,819 kg. _____

5. The UH-60 Black Hawk can carry up to 11 fully-equipped troops. _____

🎧 3-F-1

Task 2　Match the words and phrases with the definitions.

MEDEVAC (Medical Evacuation)	airframe	armed escort
CASEVAC (Casualty Evacuation)	extract	rotor blade

1. The body of an aircraft. _____

2. The long airfoil that rotates to provide the lift that supports a helicopter in the air. _____

3. Emergency evacuation of injured personnel to a hospital. _____

4. Evacuation of medical cases from one medical facility to another. _____

5. Protection for ground convoys or coastal boat traffic. _____

6. To move someone out of an area of operations. _____

Task 3　Listen and write down the primary roles of the Black Hawk.

1. (1) _____ and (2) _____ troops

2. save lives as a (3) _____ and (4) _____ platform

3. provide critical (5) _____ to (6) _____

4. deliver (7) _____ during natural disasters

5. perform as an (8) _____ and (9) _____

6. provide （10） _____ when supporting ground troops, as well as （11） _____

🎧 3-F-2

Task 4   Watch the video and order the operations that the UH-60 got involved in the history.

☐ Attack on Saddam Hussein's Forces

☐ Operation Neptune Spear

☐ Invasion of Grenada

☐ Attack in Mogadishu Somalia

☐ Invasion of Panama

Task 5   Watch again and complete the table.

Modified versions	Service it serves	Capabilities
(1) _____	(2) _____	operate off navy helicopter for aircraft carriers
Fire Hawk	US Coast Guard	(3) _____
(4) _____	(5) _____	(6) _____

Task 6   Watch again and complete the text.

It was involved in invasion of Grenada in 1983, invasion of Panama in 1989 and played a big role in the Gulf War where it participated in the largest (1) _____ in US Army history when more than 300 helicopters including Black Hawks and AH-64 Apaches attacked Saddam Hussein's forces. It wouldn't take long for several (2) _____ to arrive, offering different capabilities across the armed forces. The SH-60 Seahawk could operate off navy helicopter for aircraft carriers. The Fire Hawk can be equipped with a 1,000-gallon water tank mounted below the (3) _____, a perfect aircraft for fighting wildfires and the VH-60 White Hawk, a transport variant for the US Marine Corps, is better known as

Marine One when the president is (4) _____. The Black Hawk fought on battlefields across the world and was famously depicted in 2001's *Black Hawk Down*, a film based on the real events that occurred in Mogadishu Somalia in 1993. A mysterious (5) _____ variant of the Black Hawk also played a central role in Operation Neptune Spear, the 2011 mission that successfully killed Al-Qaida leader Osama bin Laden. The helicopter's transport ability along with its (6) _____, Hellfire missiles, the AIM-92 Stinger missiles, a six barrel minigun and a variety of (7) _____ has created a winning combination.

Today, the US military owns and (8) _____ more Black Hawks than any other air or rotor craft, with more than 2,100 of them in service. Although there are plans to (9) _____ the Black Hawk in favor of a more modern replacement, the Army has made it clear that it expects these highly capable, medium (10) _____ helicopters to remain in service through 2054, some 75 years after their introduction. For a helicopter that's built to (11) _____, it's a fitting legacy.

Task 7　Talk and present.

Work in groups and introduce the operations that the UH-60 got involved in the history, the variants of Black Hawk and its capabilities by following the outline and sentence patterns below.

Outline for reference:
- *Mission;*
- *Operations;*
- *Variants in different services;*
- *General comments.*

Sentence patterns for reference:
- *It was involved in invasion of ... and played a big role in ...*
- *A mysterious stealth variant of the Black Hawk also played a central role in ...*
- *The SH-60 Seahawk could operate ...*
- *The Fire Hawk can be equipped with ...*
- *The VH-60 White Hawk, a transport variant for ... is better known as ...*

Golf Cargo Helicopters

Scan for listening resources

A Cargo Helicopter is a helicopter that is designed or equipped for carrying infantry and/or cargo. Cargo Helicopters are typically larger and less nimble than Attack Helicopters. The largest models (e. g. the Chinook) can carry a whole Squad of infantry, while the smallest (e. g. the Little Bird) can only carry a single Fire team.

H-47 CHINOOK — A Multi-role Aircraft

The CH-47F is the latest version of the ubiquitous CH-47 Chinook. The original CH-47A reached the US Army in 1962 and became the air transport workhorse ever since. The CH-47F is an advanced multi-mission helicopter for the U. S. Army and international defense forces. It contains a fully integrated, digital cockpit management system, Common Avionics Architecture System (CAAS) Cockpit and advanced cargo-handling capabilities that complement the aircraft's mission performance and handling characteristics.

Advanced Capabilities — Block Ⅱ Features

Advanced Chinook rotor blades: A new swept-tip design increases payload by an additional 1500 pounds and reduces maintenance costs.

Improved drivetrain: More effectively takes power from the engines to the rotor blades. It provides nine percent more torque carrying capability than the current Chinook F model.

Re-engineered airframe: Expands commonality across the Chinook fleet and increases the strength of the fuselage to carry heavier equipment and payloads under higher power.

Redesigned fuel system: Instead of six individual fuel tanks, two lighter tanks will carry more fuel, reducing weight and providing for a wider mission range.

H-47 Chinook Quick Facts

The Chinook is a true multi-role, vertical-lift platform. Its primary mission is transport of troops, artillery, equipment, and fuel.

The current CH-47F/MH-47G modernization programs will ensure this tandem rotor helicopter remains in the Army fleet through the 2030s.

Chinook is the helicopter of choice for humanitarian disaster-relief operations, in missions such as transportation of relief supplies and mass evacuation of refugees.

Chinooks serve the armed forces of 19 countries around the world.

The CH-47F is operated by a crew of three, including two pilots and flight engineer.

The CH-47F design features alterations to the airframe structure to reduce the effects of vibration and other structural enhancements to the cockpit, cabin, aft section, pylon and ramp.

Task 1　Read and choose the best answer.

(　　)1. What is the primary mission of the Chinook helicopter?

　　　　A. Transporting troops.

　　　　B. Conducting artillery strikes.

　　　　C. Carrying equipment and fuel.

　　　　D. Providing humanitarian assistance.

(　　)2. What is a feature of the advanced Chinook rotor blades?

　　　　A. Increased payload capacity.

　　　　B. Improved fuel efficiency.

　　　　C. Enhanced maneuverability.

　　　　D. Reduced maintenance costs.

(　　)3. How many individual fuel tanks does the redesigned fuel system of the Chinook have?

　　　　A. 2　　　　　　B. 4　　　　　　C. 6　　　　　　D. 8

(　　)4. What does the CH-47F design feature alterations to?

　　　　A. Airframe structure.

　　　　B. Cockpit and cabin.

　　　　C. Aft section, pylon, and ramp.

　　　　D. All of the above.

(　　)5. How many countries around the world operate the Chinook helicopter?

　　　　A. 9　　　　　　B. 12　　　　　　C. 15　　　　　　4. 19

3-G-1

Task 2　Watch and choose the best title of the video.

　　A. Can the Chinook Helicopter Perform Underwater Operations?

　　B. The Reasons and Means for the Transportation Capacity of the Chinook Helicopter

　　C. Is the Chinook Helicopter the Largest Military Aircraft?

Task 3　Watch and choose the best answer.

(　　)1. What is the primary purpose of the Chinook helicopter's tandem rotor design?

　　　　A. To increase lift capacity.

　　　　B. To improve maneuverability.

　　　　C. To provide stability during hover.

　　　　D. To enable faster top speed.

(　　)2. What is the maximum weight the Chinook can carry using sling loading?

　　　　A. 7,000 pounds.　　　　　　B. 15,000 pounds.

　　　　C. 23,400 pounds.　　　　　　D. 26,000 pounds.

(　　)3. Which feature allows the Chinook to be considered semi-amphibious?

　　　　A. Its ability to hover over water.

　　　　B. Its capacity to load and unload in the water.

　　　　C. Its capability to park the back end in the water.

　　　　D. Its tandem rotor designs.

(　　)4. What is the Chinook's maximum speed?

　　　　A. 52 feet per second.

　　　　B. 23,400 miles per hour.

　　　　C. 100 miles per hour.

　　　　D. The passage does not provide this information.

(　　)5. What is the primary advantage of the Chinook's three hooks?

　　　　A. To hold the helicopter in place during takeoff and landing.

　　　　B. To allow for single-point and tandem sling loading.

　　　　C. To enable the helicopter to carry more cargo internally.

　　　　D. To provide stability during hover maneuvers.

3-G-2

Task 4 Watch and complete the table according to the example.

长度（1）___length___	（2）__98__ FT
重量（3）_____	（4）_____ pounds
最大起飞总重量（5）_____	（6）_____ LBS
引擎（7）_____	10,000（8）_____ （9）_____

Task 5 Watch and answer true（T）or false（F）.

1. Because its massive size, it flies lower than almost any other helicopter.

_____

2. This mission will be a test only for both the Chinook. _____

3. The job will be carrying an 11,000 pounds SUSV and a 30,000-pound aircraft that brings the grand total to 41,000 pounds. _____

4. Finally, the Chinook lift the heavy load. _____

5. The pilot makes the aircraft fly in an S-shape, forcing the load to go from side to side with the aircraft and stabilizing it in the intense gust. _____

Task 6 Watch and answer the questions.

1. What is the mission of Chinook in the video?

2. What does the ground crew need to do after the SUSV is hooked up?

3. What kind of challenge does Chinook have to face?

4. Why do they have such kind of training?

 3-G-3

Task 7 Watch and answer the questions.

1. What kind of operation can Chinook do as the workhorse of the US Army Aviation?

2. What are the benefits of improved equipment?

Task 8　Watch and complete.

Whether in remote locations, difficult (1) _____, high (2) _____ or hot conditions, day or night, the Chinook can do things that no other aircraft can do. As the workhorse of the US (3) _____, the Chinook is trusted to carry (4) _____ and relied upon for the most challenging battle field that (5) _____.

But over time, new and enhanced mission equipment has added weight to the aircraft, limiting what it can carry, that's why Chinook Block II is so important. It buys back lost payload (6) _____ to accommodate mission equipment that the army simply can't move today, (7) _____ tomorrow's heavy lift readiness for the US army.

These improvements expand the army's (8) _____ and the changing battlefield. The Chinook will enable soldiers to be more agile than ever before. It's the only army aircraft that can move the joint light (9) _____ vehicle with fully installed armor and protection, keeping soldiers safer and making them more effective. And it's the only army aircraft that (10) _____ the newest extended range howitzer plus ammo and crew all in one load, that minimizes the amount of time our soldiers are exposed in battle.

Task 9　Talk and present.

Work in pairs and make up a conversation to introduce the features and main tasks of the Chinook by using the sentence patterns we have learned before.

- *The CH-47 Chinook is a type of helicopter which has ...*
- *It is named for ...*
- *It provides the ability to ... to ... .*
- *Its primary roles are ...*
- *The Chinook has carried out ...*

Hotel Chinese Military Helicopters

Since its establishment in 1986, the PLA Army Aviation has witnessed great development in its weapons and equipment. In the beginning, China purchased some types of military and civilian helicopters from aboard, such as the French Dolphin helicopter. In addition, China introduced technological patents of helicopter, and purchased full sets of production facilities and main components from foreign countries. On those basis, the Z-8 helicopter and the Z-9 helicopter were developed, which became the key equipment of the PLA Army Aviation. Since the beginning of the 21st century, the Chinese aircraft industry has made major technological breakthroughs in the design and production of helicopters. A case in point is the WZ-10 attack helicopter. It is China's first domestically designed and produced attack helicopter, whose primary mission is battlefield interdiction. Today, the PLA Army Aviation has a complete and technologically advanced equipment system, which includes the WZ-19 attack helicopter, the Z-8, the Z-9 and the Z-20 utility helicopters, the Mi-8 transport helicopter, and so on.

The Harbin Z-19, also called WZ-19, is a Chinese reconnaissance/attack helicopter developed by Harbin Aircraft Manufacturing Corporation (HAMC) for the People's Liberation Army Air Force.

The Z-19 is an updated modified version of the Harbin Z-9W. It is a twin-seat tandem helicopter using commercial components from the Eurocopter AS365 Dolphin series as the Z-9 series which are licence-built versions of the Dolphin.

The Z-19 features a fenestron tail, damping its sound and therefore allows it achieve some level of acoustic stealthiness. The exhausts also protect the helicopter from infrared threats. The helicopter is installed with a millimeter wave (MMW) fire control radar. The Z-19 also features armor platings, crash resistant seats, and a turret with FLIR, TV, and laser range finder. Z-19 is also equipped with advanced helmet mounted sight (HMS), which looks different from that of WZ-10.

At the 9th Zhuhai Airshow held in November 2012, Aviation Industry

Corporation of China formally announced the official names of WZ-10 and WZ-19 at a televised news release conference, with both attack helicopters named after the nicknames of fictional characters in the Water Margin, one of the Four Great Classical Novels of Chinese literature. WZ-10 is named as Fiery Thunderbolt (Pi Li Huo), the nickname of Qin Ming, while WZ-19 is named as Black Whirlwind (Hei Xuan Feng), the nickname of Li Kui.

Task 1　Read and choose the best answer.

(　　)1. What is the primary mission of the WZ-10 attack helicopter?

      A. Battlefield interdiction　　　B. Reconnaissance

      C. Transporting troops　　　　D. Search and rescue

(　　)2. What type of helicopters were initially purchased by China for the PLA Army Aviation?

      A. WZ-10 and WZ-19

      B. Eurocopter AS365 Dolphin series

      C. French Dolphin helicopter

      D. Mi-8 transport helicopter

(　　)3. What feature of the Z-19 helicopter helps to reduce its sound and achieve some level of acoustic stealthiness?

      A. Fenestron tail

      B. Millimeter wave fire control radar

      C. Armor plating

      D. Advanced helmet mounted sight (HMS)

(　　)4. How did China obtain the technological patents and production facilities for helicopter production?

      A. They were developed domestically.

      B. They were purchased from foreign countries.

      C. They were borrowed from the Eurocopter AS365 Dolphin series.

      D. They were acquired through a joint venture with Harbin Aircraft Manufacturing Corporation.

(　　)5. Which helicopter is an updated modified version of the Harbin Z-9W?

      A. WZ-10　　　　　　　　B. Z-20

      C. Mi-8　　　　　　　　　D. Z-19

3-H-1

Task 2   Match the meanings of the following words.

1. _____ acoustic	a. relating to, producing, or employing infrared radiation
2. _____ exhaust	b. a gunner's fixed or movable enclosure in an airplane
3. _____ infrared	c. of or relating to the sense or organs of hearing, to sound, or to the science of sounds
4. _____ turret	d. of or relating to vision
5. _____ optical	e. the escape of used gas or vapor from an engine
6. _____ scout	f. lean and having prominent bone structure
7. _____ retractable	g. one sent to obtain information
8. _____ angular	h. able to be pulled backward or inside something larger

Task 3   Watch and complete.

Z-19 is a (1) _____ and (2) _____ helicopter developed and manufactured by Harbin Aircraft Industry Group for the People's Liberation Army Air Force (PLAAF). It is based on the Harbin Z-9W military (3) _____ helicopter which is a licence-built variant to the Eurocopter AS-365 Dolphin Multipurpose Helicopter.

The Z-19 can primarily conduct attack, armed reconnaissance and (4) _____ missions. It is capable of destroying enemy tanks and other land-based targets.

In May, 2010, the helicopter (5) _____ its first flight and was introduced at the China International Aviation & Aerospace Exhibition, held at the Airshow China held in Zhuhai in November, 2012. A (6) _____ helicopter designed and manufactured by the Changhe Aircraft Industry Group (CAIG) of China.

The Z-19 light attack and reconnaissance helicopter features a conventional, (7) _____ design integrating a trimmed forward fuselage and a tandem (8) _____. It also has a low acoustic signature and can (9) _____ under difficult weather and environmental conditions. A (10) _____ which has a diameter of 11.9m comes fitted into the fuselage as does a fenestron tail rotor and two stubby wings. The fuselage is also equipped with non-retractable (11) _____ for safe take-off and landing.

The helicopter has a length of 12 m and height of approximately 4.01 m. (12) _____ take-off weight is 4,500 kg, whereas the empty weight is 2,350 kg. A crew of two (13) _____ the helicopter. This includes a pilot who is seated in the front of the cockpit and a gunner (14) _____ at the rear. Crash-resistance seats and armour plates come with the cockpit to offer protection for the personnel. An infrared-suppressing exhaust system is also installed for protection against infrared-guided (15) _____.

Work in pairs and make up a conversation to introduce Z19 and compare it with Apache.

🎧 3-H-2

Task 4   Watch and answer true (T) or false (F).

1. The Z-20 helicopter is a new medium lift utility helicopter operated by China's People's Liberation Army Ground Force. _____

2. The Z-20 entered active service with the Chinese Armed Forces in 2013. _____

3. The helicopter performed its maiden flight in 2018. _____

4. The Z-20 made its first public presence at the 5th China helicopter exposition held in Tianjin in October, 2019. _____

5. The Z-20 is considered to be the country's first indigenously developed medium lift helicopter. _____

🎧 3-H-3

Task 5   Watch and choose the best answer.

(   )1. What are the improvements of the Z-20 helicopter compared with Black Hawk?

    A. Improved aerodynamic structure.

    B. A five-bladed rotor.

    C. An angular tail and two fairings.

    D. A, B and C.

（　）2. The rotor craft employs an active vibration control to _____ .

    A. minimize vibrations

    B. reduce noise

    C. improve stealth capabilities

    D. challenging weather conditions

（　）3. The multi-role helicopter was designed to _____ .

    A. operate in the highlands

    B. operate in the Plains area

    C. improve attack capabilities

    D. challenge in tropical weather conditions

（　）4. The helicopter can be deployed in multiple missions including _____ .

    A. transport of troops and cargo

    B. reconnaissance, search and rescue

    C. antisubmarine operations, assault and fire support missions

    D. A, B and C

（　）5. Which of the following statement is WRONG?

    A. The rotor craft can carry approximately 5t of cargo externally as a payload capacity.

    B. The Z-20 has the option to be armed with missiles and machine guns.

    C. Z 20's cockpit has large windows that occupy more than 50% area of the helicopters forward fuselage.

    D. It has two additional windows at the lower section to provide pilots with a wide view of the operational area.

（　）6. Which of the following statement is TRUE?

    A. The Fly by Wire Flight Control System of the helicopter is expected to deliver a maximum power of 1,600 kw.

    B. The Z-20 helicopter is powered by two locally developed WZ-10 Turboshaft engines.

    C. The Missile Approach Warning System is intended to enhance safety instability during operations.

    D. Two locally developed WZ-10 Turboshaft engines of the Z-20 helicopter is intended to reduce the burden on the pilot.

Work in pairs and make up a conversation to introduce Z-20 and compare it with Black Hawk.

3-H-4

Task 6 Watch and choose the best answer.

(   )1. What is the primary mission of the WZ-10 helicopter?

    A. Close air support

    B. Air to air combat

    C. Electronic warfare operations

    D. Cyber warfare operations

(   )2. When did the WZ-10 helicopter enter service?

    A. 2008   B. 2010   C. 2012   D. 2014

(   )3. What is one of the impressive features of the WZ-10?

    A. Advanced avionics suite

    B. Twin engine tandem seat

    C. Rugged mountainous terrain capability

    D. Urban area maneuverability

(   )4. How does the helmet mounted display (HMD) benefit the pilot?

    A. It provides real-time situational awareness and mission data.

    B. It allows the pilot to engage targets without aligning the helicopter's nose.

    C. It enhances the helicopter's maneuverability.

    D. It improves the helicopter's target acquisition capabilities.

(   )5. What is the advantage of the HMD in the WZ-10?

    A. Improved target acquisition and situational awareness

    B. Increased maneuverability in urban areas

    C. Enhanced electronic warfare capabilities

    D. Better performance in rugged mountainous terrain

Task 7 Watch and answer true (T) or false (F).

1. The WZ-10 is primarily used for transportation purposes. _____

2. The WZ-10 is a single-engine helicopter. _____

3. The WZ-10 is designed for maneuverability in urban areas only. _____

4. The WZ-10's advanced avionics suite includes a helmet mounted display. _____

5. The pilot of the WZ-10 needs to align the helicopter's nose with the target to engage it. _____

Task 8 Talk and present.

Suppose you are pilots flying the WZ-10 helicopter. Make up a short dialogue between two pilots discussing the impressive capabilities of the WZ-10 helicopter，including its ability to provide close air support，operate in any environment，and have top-notch avionics. You also praises its weapon systems and electronic warfare capabilities，which earns it worldwide recognition for its performance.

India Drones

Scan for listening resources

Military drones, or Unmanned Aerial Vehicles (UAVs), are a true game-changer in military operations. Drones offer many advantages over traditional warfare tactics with the ultimate goal of reducing soldier fatalities. Global spending on UAVs is expected to nearly reach $100 billion in the coming decade.

Why the Sudden Explosion of Military Drones

Gaining any advantage on the battlefield is fundamental to survival and winning wars. The most profound innovation in military warfare in the last several decades is the introduction of UAVs. UAVs, or drones, have powerful capabilities when it comes to surveillance and reconnaissance. Furthermore, military drones are capable of sending missiles or bombs in drone strikes. The primary advantage of UAVs is that the device does not need an onboard human pilot. Consequently, the US Military has relied upon drones for several coordinated attacks that in the past would have required more soldiers on the ground.

The Military Drone Controversy

The US government is the largest leading in military drone spending with an estimated $2.2 billion currently dedicated to UAVs. By the end of the decade, the US government currently plans to spend close to $3 billion on military drones. It's also estimated that the government has several more billions in UAV spending that do not get reported because of classified programs. The use of UAVs is skyrocketing with no end in sight. In this article, we will focus on the two types of UAVs, AAI RQ-7 and RQ-20 Puma, which are commonly used by the US Army.

1. AAI RQ-7

The AAI RQ-7 Shadow is primarily for the US Army offering reconnaissance and surveillance capabilities. The UAV is launched from a trailer-mounted catapult and requires arresting gear on an aircraft to recover the device. The impressive range of the smaller drone helps support Army operations though it is restricted in bad weather conditions.

2. RQ-20 Puma

AeroVironment RQ-20 Puma is a small, compact UAV that is battery powered. The military drone is hand-launched with a small range of a little under ten miles in distance. The compact UAV is for surveillance and intelligence gathering thanks to its infrared and electro-optical camera. Army depends on the RQ-20 Puma for surveillance. It is able to operate even in extreme weather conditions which adds value to the UAV.

Task 1   Read and choose the best answer.

(   )1. What is the primary advantage of military drones?

  A. They are capable of surveillance and reconnaissance.

  B. They can send missiles or bombs in drone strikes.

  C. They reduce soldier fatalities.

  D. They do not need an onboard human pilot.

(   )2. Which government is the largest leading in military drone spending?

  A. United States.   B. Russia.

  C. China.   D. United Kingdom.

(   )3. How is the AAI RQ-7 Shadow launched?

  A. It is hand-launched.

  B. It is launched from a trailer-mounted catapult.

  C. It is launched using arresting gear on an aircraft.

  D. It is launched from a runway.

(   )4. What is the primary purpose of the RQ-20 Puma?

  A. Surveillance and reconnaissance.

  B. Sending missiles or bombs.

  C. Intelligence gathering.

  D. Battery-powered operation.

(   )5. Which UAV is able to operate in extreme weather conditions?

  A. AAI RQ-7 Shadow.

  B. RQ-20 Puma.

  C. AAI RQ-7 and RQ-20 Puma.

  D. AeroVironment RQ-20 Puma.

3-I-1

Task 2　Watch and answer the questions.

　　1. What is the name of the drone introduced in the video?

　　2. In the video, who are trained to understand and master the art of flying these new-age surveillance assets?

　　3. Compared with other UAV models, what are the advantages of the raven?

　　4. What do the master trainers do when the teams take turns setting up and preparing to fly?

　　5. How does the instructor feel when she teaches the soldiers practice flying and landing their ravens?

Task 3　Watch again andanswer true (T) or false (F).

　　1. Many soldiers training today have already experienced with the Raven.

　　————

　　2. Flying a UAV is an exciting experience for many of the operators. ————

　　3. Operators take instruction of flight from the ground. ————

　　4. The soldiers alternated the roles of vehicle operator and mission operator.

　　————

　　5. Instruction on fundamentals of flight is classified, and instructors don't usually tell soldiers how to operate them. ————

Task 4　Work in pairs. Describe the RQ-11 Raven according to the specifications given below.

RQ-11 Raven　Specifications			
Builder	AeroVironment Inc	**Cruising speed**	95 km/h (28–60 mph)
Role	battlefield reconnaissance	**Range**	10 km (6.2 miles)
Wing Span	55 inches (4 ft 3 in)	**Service Ceiling**	4500 m (15,000 ft)
Length	36 inches (3 ft 7 in)	**Flight Time**	60–80 Minutes
Height	4,500 m (15,000 ft)	**No. In Inventory**	
Weight	1.9 kg (4.2 lb)	**Payloads**	interchangeable: optical, infrared, and IR cameras

Engine	Aveox 27/26/7-AV electric motor	Cost	$ 35,000
Maximum speed	56 km/h	User Countries	United States

 3-I-2

Task 5 **Watch and complete.**

One of the key new things now is the new （1） _____ , with a total different design it is based on a much more （2） _____ approach it has a new suite of （3） _____ and it has also increased capabilities. There is a whole range of them, one of them is a GPS （4） _____ capability which not only increases the system's （5） _____ , but also opens up for other types of （6） _____ than we have been doing before. As an example to be able to fly （7） _____ which is important for many （8） _____ customers out there. Cameras are of the （9） _____ generation, so that also the （10） _____ of the video stream. And the pictures, which the （11） _____ will get here, is significantly （12） _____ than it was on the last generation. The whole system is approximately the same （13） _____ as before, but the air vehicle the helicopters are a little bit （14） _____ , but they still fit into the same （15） _____ as we have been using now for some years.

 3-I-3

Task 6 **Watch and fill the missing information.**

Black Hornet	
advantages	（1） _____
launch time	（2） _____ seconds
flight endurance	（3） _____ minutes
speed	（4） _____ kilometers per hour
range	（5） _____ kilometers

3-1-4

Task 7　Watch and choose the best answer.

(　　)1. What methods did the US use to locate Qasem Soleimani?

    A. Israeli Intelligence and classified informants.

    B. Classified informants and electronic intercepts.

    C. Reconnaissance aircraft and other surveillance.

    D. White House orders and drone strikes.

(　　)2. What is the range of the MQ-9 Reaper drone?

    A. Approximately 570 kilometers.

    B. Less than 480 kilometers per hour.

    C. Over 1,800 kilometers.

    D. At least more than an hour and a half.

(　　)3. Why did Israeli military mobilize on Friday?

    A. In response to an Iranian threat.

    B. Due to a specific reason.

    C. To assist in finding Qasem Soleimani.

    D. As a precautionary measure.

(　　)4. Where did the US drones most likely take off from?

    A. Ali Al Salem Air Base in Kuwait.

    B. Al Udeid Air Base in Qatar.

    C. Al Fora Air Base in the UAE.

    D. Baghdad International Airport in Iraq.

(　　)5. Why were tensions between Israel and Iran already high before Soleimani's assassination?

    A. A failed Mossad assassination plot.

    B. Israeli military mobilization.

    C. Memorial visit by Soleimani.

    D. Iranian response expected.

Task 8　Watch again and answer true (T) or false (F).

1. The US missile strike targeted Qasem Soleimani and Abu Mohdi al Mohandas. _____

2. The US found Qasem Soleimani through classified informants, electronic intercepts, reconnaissance aircraft, and other surveillance. _____

3. The drone strike took place as the cars left Baghdad International Airport. _____

4. There are operational US drone bases in the area where the strike took place. _____

5. Mossad was involved in the drone strike that killed Qasem Soleimani. _____

3-1-5

Task 9 Watch and answer the questions.

1. Why did General Qasem Soleimani wear a hat and mask to conceal his identity during the flight?

2. What role did the US Reaper drones play in the operation to eliminate Soleimani?

3. How did the CIA prepare for the operation to kill Qasem Soleimani?

4. How did the Reaper drones identify their target before firing the missiles that killed Soleimani and others?

5. What impact did the drone strike have on Soleimani and his entourage at the airport?

Task 10 Watch again and answer true (T) or false (F).

1. When Soleimani arrived at Baghdad International Airport, there were three British Reaper drones hovering silently over the area. _____

2. Late on New Year's Eve 2019, President Barack Obama decided to take the risk and ordered a drone strike that killed Soleimani. _____

3. At 10 : 30 a.m., three Reaper drones took position over Baghdad International Airport. _____

4. Although Solomani's flight had been delayed earlier, the CIA was prepared for every contingency in this operation. _____

5. The Reaper drones were not armed during the operation. _____

6. When the Reapers fired three missiles, two were directed at the lead vehicle and one at the other. _____

Task 11 Talk and present.

Work in pairs. Talk about a drone you know. What are its advantages and disadvantages compared with other drones?

Juliet Chinese Military Drones

Scan for listening resources

Unmanned aerial vehicles (UAVs) are a vital part of China's strategy for winning "intelligent conflicts". A wide range of military drones enables the People's Liberation Army (PLA) to perform intelligence, reconnaissance as well as direct attacks on enemy targets at relatively low cost.

The WZ-7 Soar Dragon reconnaissance drones (High-Altitude Long Endurance type, HALE) are capable of operating at altitudes above 15,000 m (around 50,000 ft) and intended mainly for intelligence and reconnaissance missions. WZ-8, was presented during a military parade on Oct. 1, 2019. It's said to be able to achieve supersonic speed and has stealth technology, making it less visible to radars and more difficult to be detected by the enemy. WZ-7 and WZ-8 are expected to make up the satellites' observation capabilities, which are limited by their orbital trajectories.

In 2020, work on CR500 Golden Eagle unmanned helicopter was completed by the unit whose task would be to support armored forces, self-propelled artillery and infantry. It is supposed to facilitate locating enemy units by tanks and artillery as well as assess the inflicted damage.

Furthermore, the Chinese unmanned aerial vehicles fleet includes medium-altitude long-endurance (MALE) units. The newer model is capable of carrying twelve air-to-surface missiles including: Lan Jian 7 (Blue Arrow 7), TG100 and AR-1/HJ-10 using laser guidance/GPS. It can reach a speed of up to 370 km/h (230 mph) and a height of 9,000 m (around 30,000 ft). GJ-2/Wing-Loong II is expected to be a cheaper alternative to American models, and the CH-4 models is available in reconnaissance as well as mixed attack and reconnaissance variants. This model, similarly to the Wing Loong series, is recognized as the cheaper equivalent to the American MQ-9 Reaper.

Unmanned aerial vehicles are still a component which supports the traditional armed forces, but their role is steadily growing, mainly in the area of improving situational awareness. The digitalization of the battlefield will increase the importance of the signals intelligence (SIGINT) capability involving the interception of electronic communication or tracking information flows. As this sector develops, land, air and naval units are expected to be widely equipped with various unmanned systems, serving reconnaissance, logistic and offensive purposes.

Task 1　Read and choose the best answer.

(　)1. The WZ-7 Soar Dragon reconnaissance drones are primarily used for which type of missions?

A. Direct attacks on enemy targets.

B. Supersonic speed capabilities.

C. Support of armored forces.

D. Intelligence and reconnaissance.

(　)2. What advantage does the WZ-8 drone have over other drones?

A. It can achieve supersonic speed.

B. It has stealth technology.

C. It is less visible to radars.

D. It has a longer flight endurance.

(　)3. What is the purpose of the CR500 Golden Eagle unmanned helicopter?

A. Intelligence and reconnaissance.

B. Support of armored forces.

C. Self-propelled artillery.

D. Inflicting damage on enemy units.

(　)4. Which unmanned aerial vehicle model is recognized as a cheaper alternative to American models?

A. GJ-2/Wing-Loong Ⅱ　　　　B. WZ-8

C. CH-4　　　　D. Lan Jian 7

(　)5. What role do unmanned aerial vehicles play in the traditional armed forces?

A. Direct attacks on enemy targets.

B. Intelligence and reconnaissance.

C. Improving situational awareness.

D. Logistic and offensive purposes.

Task 2   Match the units of measurement.

1 foot   1,852 meters per hour

1 knot   0.3048 meters

1 mile   0.45359237038038 kg

1 ib   1.609344 km/h

1 mph   1.609344 kilometers

Task 3   Watch and fill the missing words.

The WING LOONG Ⅱ unmanned aerial vehicle ... Intended for （1） _____,
（2） _____ and （3） _____. The drone can carry a maximum external
（4） _____ of 480 kilograms. It can carry up to 12 （5） _____ bombs or
missiles, including FT-9/50 bombs, GB3 250 kilogram laser-guided bombs and
TL-10 missiles. The medium altitude long （6） _____ unmanned aircraft can
endure for 32 hours and cruise at a maximum speed of （7） _____. It has a
maximum takeoff weight of 4,200 kilograms, and a service ceiling of
（8） _____.

**Task 4   Work in pairs. Student A asks Student B questions and completes the
specifications.**

The WING LOONG Ⅱ	
Payload	_____ kilograms
Armament	_____ Laser-guided bombs or missiles
Maximum speed	_____ knots
Maximum takeoff weight	_____ kilograms
service ceiling	_____ feet

Task 5   Watch and answer the questions.

1. What's the other name of CH-5 UAV?

2. What is CH-5 used for?

3. What can the hard points on each wing be attached with?

Task 6   Work in pairs. Student A asks Student B questions and completes the specifications.

CH-5 UAV	
Payload	_____ kilograms
Armament	_____ guided anti-Armour missiles
	and _____ missiles
Maximum speed	_____ knots
Maximum takeoff weight	_____ kilograms
Engine	_____ horsepower
Durance	_____ hours

Task 7   Talk and present.

Discuss what characteristics an excellent drone should have.

Kilo Further Reading

◇ **Text A**

Why Army Helicopters Have Native American Names?

1. In 1962, based on the existing system of the Air Force, the US military completely banned the old aviation naming system of the Navy, Marine Corps and Army, and gradually formed today's naming standard. The standard covers aircraft, helicopters, missiles, drones, airships and all equipped military aircraft. These military aerospace vehicle designations are also known as "MDS designations" (Mission-Design-Series designations), which consists of three parts and can be used to interpret all military aircraft from the model designation.

2. Take the AH-64 Apache attack helicopter as an instance, "A" for ground attack, "H" for helicopter, and 64 represents the independent design number during its equipment development. However, it is his nickname "Apache" that is more well-known and widely used.

3. You may have noticed there's a pattern in US Army helicopter names — Apache, Black Hawk, Chinook, Iroquois, and others. These crucial aircraft are all named after Native American tribes or figures.

4. But have you ever wondered why?

5. The US military has a long history with Native Americans. Armed conflicts between the two were commonly known as the American Indian Wars and were fought intermittently from the time the US was first settled by Europeans to early in the 20th century. In the 100 years of the 19th century, the US government and military launched more than 200 surprise attacks and sweeps on various Indian tribes. But Native Americans also served as some of the fiercest fighters for the United States in these years. The free and brave Indian tribes did not sit still but fought heroically with the most primitive weapons, and some tribes even quickly learned to use weapons like muskets. Indian heroes such as Black Hawk, Apache and Chinook appeared, which were both tribal names and personal names for Indian warriors. The Americans surely have hatred for Indians, but they also have deep respect and hidden fears for them. In fact, 32 Native Americans have earned the nation's highest military award, the Medal of

Honor.

6. The tradition of naming helicopters after Native Americans was once an official regulation. That regulation no longer stands, but the tradition continues.

7. How it all came about

8. According to an unnamed Army museum official, the naming convention goes back to before the Air Force split from the Army in 1947 when Army Gen. Hamilton Howze was assigned to Army Aviation. His mission was to develop doctrine and the way forward when it came to employing Army aircraft and how they would support warfighters on the ground.

9. According to the museum official, Howze wasn't a fan of the names of the first two helicopters — Hoverfly and Dragonfly. So, he laid out instructions for naming the helicopters after their abilities.

10. Howze said since the choppers were fast and agile, they would attack enemy flanks and fade away, similar to the way the tribes on the Great Plains fought during the aforementioned American Indian Wars. He decided the next helicopter produced — the well-known H-13 of "M. A. S. H." fame — would be called the Sioux in honor of the Native Americans who fought Army Soldiers in the Sioux Wars and defeated the 7th Calvary Regiment at the Battle of Little Bighorn.

11. That's likely how Army Regulation 70-28 was created in 1969. The regulation listed criteria on how popular names would be given to major items of equipment.

12. Name choices had to:

* Appeal to the imagination without sacrificing dignity.

* Suggest an aggressive spirit and confidence in the item's capabilities.

* Reflect the item's characteristics including mobility, agility, flexibility, firepower and endurance.

* Be based on tactical application, not source or method of manufacture.

* Be associated with the preceding qualities and criteria if a person's name is proposed.

13. According to AR 70-28, Army aircraft were specifically categorized as requiring "Indian terms and names of American Indian tribes and chiefs." Names to choose from were provided by the Bureau of Indian Affairs, which were generally one of five. Arms were named with Indian pronunciation, considering the history and aircraft missions. The name should be able to arouse people's

inspiration of bravery, and show the enterprising spirit and confidence in the capability of the aircraft.

14. Other categories included tanks, which were to be named after American generals like Gen. William Tecumseh Sherman; infantry weapons would receive names for famous early American pioneers like Daniel Boone and Davy Crockett; and assault weapons would get fearsome reptile and insect names like cobra and scorpion.

15. AR 70-28 was eventually rescinded and replaced with policies that didn't mention that criteria, but it's clear that the tradition has continued. You only have to look back to 2012 when the Army named its current primary training helicopter, the UH-72 Lakota, after the Lakota tribe of the Great Sioux Nation in North and South Dakota.

16. On June 10, 2012, Lakota elders ritually blessed two new South Dakota Army National Guard UH-72A Lakotas at a traditional ceremony on the Standing Rock Reservation in North Dakota. Ceremonies like these happened often over the past several decades.

17. Therefore, the name of a helicopter conveys not only its model and function, but also the cultural and spiritual values in it.

Task 1 Read and match the meanings of the following words.

1. _____ aerospace	a. moving quickly and lightly
2. _____ surprise attack	b. the industry of building aircraft and vehicles and equipment to be sent into space
3. _____ sweep	c. a poisonous African or Asian snake that spreads out the skin on its neck when it is angry
4. _____ agile	d. a search of an area of ground or sea
5. _____ flexibility	e. an attack without warning
6. _____ cobra	f. a large group of related families who live in the same area and share a common language, religion, and customs
7. _____ scorpion	g. the ability to make changes or to deal with a situation that is changing
8. _____ tribe	h. an animal like a large insect with a long curved tail containing a poisonous sting

Task 2 Read and answer true (T) or false (F).

1. Black Hawk is a nickname for UH-60 Medium Utility Helicopter named after the animal hawk. _____

2. The US Air Force has a long history of more than 100 years. _____

3. The Native Americans have never earned American's highest military award, the Medal of Honor. _____

4. According to the Army Regulation 70-28, names of helicopters should be based on tactical application, not source or method of manufacture. _____

5. The tradition of naming helicopters after Native Americans is an official regulation today. _____

Task 3 Read and answer the questions.

1. What is the modern naming system of the American aircraft?

2. Can you interpret the name of AH-64 Apache attack helicopter?

3. Who is said to start the convention of naming American Army helicopters using Native American names? And what is his mission?

4. What are the rules of Army Regulation 70-28 listed on how popular names would be given to major items of equipment?

5. What kind of spirit does the Native American names evoke in helicopters?

Task 4 Translate the following sentences into Chinese.

1. The US military has a long history with Native Americans. Armed conflicts between the two were commonly known as the American Indian Wars and were fought intermittently from the time the US was first settled by Europeans to early in the 20th century.

2. According to an unnamed Army museum official, the naming convention goes back to before the Air Force split from the Army in 1947 when Army Gen. Hamilton Howze was assigned to ArmyAviation.

3. Arms were named with Indian pronunciation, considering the history and aircraft missions. The name should be able to arouse people's inspiration of bravery, and show the enterprising spirit and confidence in the capability of the aircraft.

A Sociology of the Drone

(excerpt)

Defence politicians, international law experts, philosophers and not least the critical public — all of them have increasingly led debates for years on the topics of the acquisition and employment of combat drones. The reason why sociological aspects of technology stand back behind ethical and jurisdictional questions in the current debates of combat drones may be due to the fact that the latter questions demand an answer more pressingly. By contrast, sociology studies the reasons for the development of technology and their long-term effects on social action, considering a longer background of time. Sociologists can enrich the drone discourse by giving answers to the questions why societies are technologically at the point they are and in which way combat drones change soldiers and local population, armed forces, as well as states and societies.

Regarding the development and employment of combat drones, the question that arises is *Why*. Why do military drones exist? Why are drones equipped and employed in a particular way? Approaching this question from a macro-perspective, combat drones may appear as another expression of western societies' estrangement from war. The drone as a standoff weapon keeps off the war from the post-heroic, casualty-sensitive societies that shy away from putting their own soldiers at the risk of being killed.

Another macro-perspective on the development and employment of drones is that of the social cost pertaining to wars. The argument goes, that large-scale wars are too cost-intensive for globalized and networked societies. Therefore, conflicts are rather being downscaled to small-scale wars. Combat drones area perfect means for small-scale, but often endless wars.

In addition to the focus on innovation processes, another perspective asks about the diffusion of ideas and technologies. More than 70 states own drones and more than 50 states are developing their own drones. A swift diffusion of military drones can be found among other states' armed forces. The adoption of drone technology is not only triggered by aspects of effectiveness. For example, organisational prestige plays a role here. There are indications that states acquire drones in order to demonstrate the modernity of their own military. The statements of the former German Federal Minister of Defence, Thomas de

Maizière, exemplify this for Germany. During a debate in the plenum of the German Parliament, de Maizière lists a series of effectiveness-based arguments in favour of drone acquisition. But he also says: Germany has to be part of this future technology. We cannot say we are fine with the pony express while everybody else is developing the railway. This is not possible.

The sociology of technology not only asks about the conditions for the development of technology but also about the consequences of the introduction for societies, organisations and people. Remaining at first at the microsociological level, the questions arise, whether and how the existence of combat drones influences individuals in their behaviour.

The existence and the use of combat drones affects and changes the life and the identity of people. A group of people who are directly affected by the employment of combat drones are those inhabiting the areas where combat drones are employed. A study interviewed the witnesses and victims of drone attacks, which confirms that the existence of combat drones has changed the residents' everyday life beyond causing physical and psychological pain: The practice of the double strike for example, i.e. the launch of a second attack after the first one, led to the members of village communities no longer approaching the wounded people to help. Furthermore, some people refrain from meeting in groups, as this could attract the attention of the all-monitoring drones and their operators. However, the assemblies of the elders are a traditional form of dispute settlement in this region. Other social practices—children's school attendance, participation in funerals—are affected, according to the authors of the study.

However, the existence of drones also affects the lives of the drone operators. First, the psychological issue of the impact of posttraumatic stress disorder on drone operators has come to the focus of social science research. Meanwhile, from a military sociological perspective, it is also interesting how the new professional profile of the drone operator will integrate into the already existing social framework of different status groups within the military. Drone operators in the US Armed Forces are counted among the group of combat aircraft pilots. But is the job of a drone operator comparable to that of an airplane pilot? Is the workload comparable? Many "real" pilots doubt this and it is questionable whether drone operators will ever have a similar institutional acknowledgment as their colleagues.

Further phenomena interesting to organisational sociology are the impact of the existence of combat drones and other standoff weapons on the composition, equipment and therefore the capability profile of the military. Drones are currently the highest developmental form of a military strategy orientation of armed services who wish to minimise their footprint in the operational theatre.

Ultimately, the current drone discourse reflects and intensifies the general discourse of the western world about war and violence. One may ask whether the drone discourse, which reflects in its ethics dimension about How wars are conducted — as just and legitimate as possible — is not yet another side track leading away from the real urgent question: How can the causes of certain conflicts be removed and thus wars and violent collective disputes be ended and prevented?

Task 5 Translate the following sentences into Chinese.

1. Sociologists can enrich the drone discourse by giving answers to the questions why societies are technologically at the point they are and in which way combat drones change soldiers and local population, armed forces, as well as states and societies.

2. Drones are currently the highest developmental form of a military strategy orientation of armed services who wish to minimise their footprint in the operational theatre.

Chapter Four

What We Do:
Mission and Command

Alpha Overview

Army Aviation get the chance to fly some of the most advanced aircraft in the world all while playing a key role in high-stakes aviation missions.

1. Army Aviation conducts Air-ground operations as the aerial maneuver force of the combined arms team, or as an independent maneuver force in support of ground forces conducting offensive, defensive, stability, and Defense Support of Civil Authorities operations. Regardless of the type of mission performed by the ground force, most aviation operations are offensive in nature and designed to provide an asymmetric advantage. Aviation operations are most effective when assets are task organized to correctly support the higher headquarters mission.

Unique Capabilities

2. Aviation is not a substitute for any other member of the combined arms team. Rather, it brings a degree of versatility not replaced by other members of the combined arms team and a range of unique capabilities that complement those of the other combat arms.

3. Aviation maneuvers rapidly and simultaneously in the ground commander's battlespace to bring decisive combat power at the decisive points and times in the area of operations (AO). There is an inseparable linkage between maneuver and fires. Army Aviation maneuvers while leveraging organic firepower to shape the battlespace or conduct decisive operations as directed by the force commander.

4. Synchronizing aviation maneuver with ground maneuver — by enhancing reconnaissance, providing security, and conducting attacks and counterattacks — allows the friendly force commander to shape the battlespace to set the conditions for the close fight and achieve a positional advantage in both time and space. Linked with deep fires, aviation maneuver offers the ground commander the capability to influence events simultaneously throughout his AO.

Missions

5. Aviation units operate in the ground regime. As a fully integrated member of the combined arms team, aviation units conduct combat, CS, and CSS operations. Aviation units operate across the entire length and breadth of the AO (close, deep, and rear), and can be expected to conduct simultaneous operations, 24 hours a day.

6. Aviation combat missions are performed by maneuver forces engaged in shaping the battlespace and conducting decisive combat operations by employing direct fire and standoff precision weapons in combined arms operations.

7. Aviation combat support (CS) is the operational support and sustainment provided to forces in combat by aviation units, that is to support combat elements in contact with the enemy.

8. Aviation combat service support (CSS) is the assistance provided by aviation forces to sustain combat forces. Army Aviation provides air movement of personnel, equipment, and supplies; and performs aeromedical evacuation and aviation maintenance.

Task 1 Read and answer true (T) or false (F).

1. Army Aviation can conduct various operations but aviation operations are typically defensive. _____

2. Aviation can not be replaced by any other member of the combined arms team. _____

3. Army Aviation can both maneuver and provide organic firepower at the same time. _____

4. Army Aviation units conduct combat, CS, and CSS operations. _____

5. Aviation combat service support is the assistance provided to forces in combat. _____

Task 2 Compare Aviation combat, CS, and CSS operations and identify the differences among them.

Task 3 Read the following passage and identify which types of operations (combat, CS, CSS) Air Movement, Air Assault and Aerial sustainment respectively belong to and then offer reasons according to the definitions given in the passage.

Air movement operations are conducted to reposition units, personnel, supplies, equipment, and other critical combat elements in support of current and/or future operations. Air movement operations allow the ground force commander to control the tempo of operations and meet the enemy force at the time and place of choice as he or she sets conditions. Utility and cargo helicopters supplement ground transportation to help sustain continuous offensive and defensive operations, and allow the supported commander to overcome difficult terrain and time constraints on operations.

Air assault is the movement of friendly assault forces by Rotary Wing aircraft to engage and destroy enemy forces or to seize and hold key terrain. Air assaults extend the tactical and operational reach of the combined arms team by overcoming the effects of terrain, achieving surprise, and isolating, dislocating, or destroying enemy forces by rapidly massing combat power at the maneuver commander's time and place of choice.

Aerial sustainment is the movement of equipment, material, supplies, and personnel by utility, cargo, and fixed-wing assets for operations other than air assault and combat support. Missions include intratheater airlift, administrative relocation of troops and nonmilitary personnel, and administrative relocation of equipment, material, and supplies.

Task 4   Match the missions with the types of operations.

Operations	Aviation Missions
Combat	
Combat Support	
Combat Service Support	

1. Attack
2. Air Assault
3. Reconnaissance
4. Movement to Contact
5. Security
6. Air Movement
7. Aeromedical Evacuation
8. Personnel Recovery
9. Command and Control Support

Bravo Radio Communications

Scan for listening resources

Radio communications are a critical link which can be a strong bond between pilot and controller. The most important thought in pilot—controller communications is understanding. It is essential, therefore, that pilots acknowledge each radio communication by using the appropriate aircraft call sign. Brevity is important, and contacts should be kept as brief as possible.

Radio Technique

1. Listen before you transmit. Many times you can get the information you want by monitoring the frequency.

2. Think before keying your transmitter. Know what you want to say and whether it is lengthy.

3. The microphone should be very close to your lips and after pressing the button, a slight pause may be necessary to be sure the first word is transmitted. Speak in a normal, conversational tone.

4. Be alert to the sounds or the lack of sounds in your receiver. Check your volume and recheck your frequency.

5. Be sure that you are within the performance range of your radio equipment and the ground station equipment.

Radio Protocol Best Practices

1. Identify with whom you want to communicate by using their call sign.

2. Pause a moment after pressing the "push-to-talk" (PTT) button.

3. Be direct and short when communicating.

4. Speak slowly and clearly.

5. Spell out letters and numbers, using the Military Alphabet (NATO Phonetic Alphabet).

6. Use correct lingo and prowords to reduce confusion and shorten transmitted messages.

Contact Procedures

1. Initial Contact

The terms *initial contact* means the first radio call you make to a given facility or the first call to a different controller within a facility. Use the following format:

Name of the facility being called;

Your full aircraft identification as filed in the flight plan or as discussed in Aircraft Call Signs;

The type of message to follow or your request if it is short;

The word "Over" if required.

2. Subsequent Contacts and Responses

Use the same format as used for the initial contact. If the situation demands your response, take appropriate action or immediately advise the facility of any problem. Acknowledge with your aircraft identification, either at the beginning or at the end of your transmission, and one of the words "Wilco," "Roger," "Affirmative," "Negative".

Precautions in the Use of Call Signs

Improper use of call signs can result in pilots executing a clearance intended for another aircraft. Call signs should never be abbreviated on an initial contact or at any time when other aircraft call signs have similar numbers/sounds or identical letters/number.

Task 1   Read out and repeat the military alphabet.

A: Alpha	B: Bravo	C: Charlie	D: Delta
E: Echo	F: Foxtrot	G: Golf	H: Hotel
I: India	J: Juliet	K: Kilo	L: Lima
M: Mike	N: November	O: Oscar	P: Papa
Q: Quebec	R: Romeo	S: Sierra	T: Tango
U: Uniform	V: Victor	W: Whiskey	X: X-ray
Y: Yankee	Z: Zulu		

4-B-1

Task 2   Listen and write down the call signs.

1. _____

2. _____

3. _____

4. _____

 4-B-2

Task 3   Listen and write down the prowords and then match the prowords with the definitions.

1.	A. What is my signal strength? Can you hear and understand my transmission?
2.	B. This transmission is from the station whose call sign immediately follows.
3.	C. This is the end of my current transmission and I am now expecting a response.
4.	D. I have received your last transmission.
5.	E. This is the end of my transmission to you and no answer is required or expected.
6.	F. I'm moving in a vehicle from [place].
7.	G. I'm moving in a vehicle and I'm going to a place.
8.	H. I'm getting out of the vehicle. I won't be in radio contact.
9.	I. A message which requires recording is about to follow. Transmitted immediately after the call.
10.	J. Initial call received, continue with the rest of your message.
11.	K. Confirm you have received and understood a message.
12.	L. I have received your message, understand it and will comply.

13.	M. Yes/Correct.
14.	N. No/Incorrect.
15.	O. I have a question.
16.	P. I shall spell the next word phonetically.
17.	Q. I have not understood your transmission, please repeat.
18.	R. I am repeating a previous transmission or portion indicated.
19.	S. Your transmission is at too fast a speed. Reduce speed of transmission.
20.	T. Repeat this transmission back to me.
21.	U. The following is my response to your instructions to read back.
22.	V. I hereby indicate the separation of the text from other portions of the message.
23.	W. Your transmission has been received clearly.
24.	X. Your transmission has been received, but cannot be understood.

 4-B-3

Task 4　Listen and complete.

Conversation 1

　　A R1. This is R2. (1) _____. Over.

　　B R1. Loud and clear. (2) _____.

　　A R2. OK. (3) _____.

Conversation 2

　　A R1. This is R2. (4) _____ Hill 102 en route to Hill 201. Over.

　　B R1. (5) _____. Over.

　　A R2. I say again. Mobile from Hill 102, heading to Hill 201. (6) _____.
Over.

B R1. (7) _____. Out.

Conversation 3

    A R1. This is R2. (8) _____ for 20 minutes at Hill 102. Over.

    B R1. Roger. Out.

Conversation 4

    A R1. This is R2. (9) _____. Over.

    B R1. Roger. Out.

Work in pairs and practise the conversations.

🎧 4-B-4 & 4-B-5

Task 5   **Listen and complete the conversations.**

Conversation 1

    A H2. This is H1. (1) _____. Over.

    B H2. Send. Over.

    A H1. Enemy spotted 2 km west of current position. (2) _____. Move to Delta Shack, 2 km east, for cover. (3) _____. Over.

    B H2. I read back: enemy 2 km west, moving to Delta Shack east 2 km for cover. Over.

    A H1. (4) _____. Over.

    B H2. (5) _____. Over.

    A H1. Roger. Out.

Conversation 2

    A H2. This is H1. Message. Over.

    B H2. (1) _____. Over.

    A H1. An explosion reported in residential area at (2) _____. One soldier (3) _____. Request immediate medevac. Over.

    B H2. (4) _____. Is the area safe? Over.

    A H1. (5) _____. Area secure, no hostiles nearby. (6) _____. Over.

    B H2. We are sending an (7) _____. Stay at the site. Over.

    A H1. Roger. Out.

Work in pairs and practise the conversations.

Task 6   Watch and answer the questions.

1. What do the soldiers in the helicopter do when they find civilians are under fire?

2. What is the ROE mentioned in the video clip?

Task 7   Listen again and complete.

A Chief，we got (1) _____ here at nine o'clock.

B I got it，Matt. I don't think we can (2) _____ this.

B (3) _____, Super 64. We got militia shooting civilians at the food distribution center. (4) _____.

C Super 64，are you taking fire? (5) _____.

B (6) _____, command.

C UN's jurisdiction，64. We cannot (7) _____. Return to base. Over.

B (8) _____.

Work in pairs and practise the conversations.

Task 8   Talk and present.

Suppose you are required to conduct a patrol/reconnaissance task and you should report what you hear and see in the scene and then request permission for support/further action. Make a conversation with the prowords and sentence patterns you've learnt in this lesson.

Charlie Attack

Scan for listening resources

Army Aviation attack or reconnaissance units, employing Manned-Unmanned Teaming (MUM-T), conduct attacks in support of offensive, defensive and stability operations throughout the depth of the AO. This is done either as a decisive or shaping operation in support of ground forces. Decisive operations aim to find, fix, and destroy enemy forces (especially moving forces), and to confirm intelligence. Shaping operations establish conditions for success of the decisive operation by setting the battlefield to our advantage.

Army Aviation conducts attacks at multiple echelons. These can range from elements as small as attack teams using MUM-T, or a single armed UAS, up to battalion or squadron level.

Army Aviation attacks are typically supported with integrated joint fires. The methods of employment are solely driven by whether a friendly ground maneuver force is in direct contact with the targeted enemy force or not; this factor determines who controls the aviation maneuver and fires. Regardless of the methods employed, the tactical task assigned to the attack or reconnaissance unit is attack to destroy, defeat, disrupt, divert, or delay.

To ensure success, Army Aviation attacks with the necessary combat power, tempo and intensity to overwhelm the enemy force. Audacity, speed, concentration of combat power at the right time and place, violence of execution, simultaneity of joint fires with ground and air maneuver, and maximizing the element of surprise are all essential components of successful Army Aviation attacks.

Attacks are triggered based on enemy events, time, friendly actions or a combination. During execution, the Army Aviation attack or reconnaissance units use maneuver and direct and/or indirect joint fires to place the enemy in a position of disadvantage. To achieve superiority over the enemy, the commander must take advantage of the range, precision, and lethality of all available fires. He or she also must gain and maintain information superiority through in-depth reconnaissance and continuous maneuver to positions of advantage using speed, maneuverability, maximum standoff, and the available terrain.

Task 1 Read and answer true (T) or false (F).

1. Army Aviation units conduct attacks in support of shaping, offensive, defensive and stability operations throughout the depth of the AO. _____

2. All levels of Army Aviation units, from teams to groups can conduct attacks. _____

3. Army Aviation attacks were generally conducted by the attack helicopters on their own. _____

4. A successful Army Aviation attack is based on the audacity, speed, concentration of combat power. _____

5. During execution, the Army Aviation unit should maintain superiority over the enemy in position, information and fires. _____

Task 2 Read and answer the questions.

1. What are the major weapons used in an Army Aviation attack?

2. What is the fundamental role played by the Army Aviation units?

3. Can you find the respective synonym for"tempo" and "intensity" in the text?

4. To use firepower effectively, what are the elements that should be considered?

 4-C-1

Task 3 Listen and answer the questions.

1. What was the coalition war plan?

2. Why was the first day of air assault the most risky?

3. What was the priority for day one of air campaign?

4. What and when was the H Hour?

5. What was the flight formation of aviation engaging the ground targets?

6. What was the call sign and job of this task force?

4-C-2

Task 4   Listen and complete.

The two groups of Apaches (1) _____ near two radar sites 30 seconds early and come to (2) _____ in lines of four abreast. As they (3) _____ precisely 2:38 am, they (4) _____ the compounds through their infrared cameras. At first, it appears that the Iraqis are oblivious to their presence. But with 10 seconds to go, the compounds' lights go out, and figures are observed running around outside.

At 2:38 am H hour, Desert Storm begins. The Apaches (5) _____ a furious rain of destruction on the radar installations. Hellfire missiles are first (6) _____ power generators. Further Hellfires and 70-millimetre rockets are (7) _____ radar dishes, (8) _____ vehicles and communications antennas.

As the attack helicopters moving closer, they (9) _____ everything they have on the site. Such is the importance to (10) _____ the radar, the Apache crews then move in closer still and (11) _____ whatever is left standing with 30-millimetre cannon. An enormous explosion ends the attack when an ammunitions (12) _____ is hit by cannon fire.

During the 4 minutes of (13) _____, 27 Hellfires, 100 rockets and 4,000 cannon round shave been unleashed on the sites. The crew's reporting "Total destruction of the two sites for no loss". They turn back to the (14) _____.

 4-C-3

Task 5   Listen and complete.

August 1990, Iraqi dictator Saddam Hussein sends his army into neighboring Kuwait. Its oil fields are rich for the taking and Hussein may have his eye on Saudi oil too. Saudi Arabia quickly gathers its allies. The 39-nation (1) _____ is the largest since World War II. They vowed to bring Hussein to his knees. "Withdraw from Kuwait or face a coalition ready and willing to (2) _____ all means necessary."

Eight Apaches from the 101 Airborne Division prepare for the 200 Mile flight into Iraq. "We had to take out the radar (3) _____ at 2:38 in the morning, so all the aircraft could fly through this (4) _____." The Apaches take off and speed across the desert, (5) _____ just 50 feet above the ground to avoid radar

(6) _____ . There's a total blackout on (7) _____ lights, total radio silence. If Hussein's radar detects them, they risk being shot down by the very target they've come to kill. It takes just over 2 hours for the Apaches to reach the radar towers. When they're 5 miles away, they (8) _____ for a moment (9) _____ their target. This is actual footage from General Cody's mission, the very first Apache strikes of Desert Storm. When we pulled the trigger, all four aircraft shot at the same time, you got 4 to 6 Hellfires en route. The time of flight's about 12 seconds. And then all of a sudden they started hitting. And when they started hitting, things started (10) _____ pretty quickly, especially when we hit where the (11) _____ and all the fuel was. At about 4 kilometers, we started opening up with the (12) _____ that had the Flechette (13) _____ and that was to take out the air defense guns that were out in front. And then as we get closer, we open up in the 30 millimeter and finish off the job. It's 4 minutes of mayhem with 100% destruction. We've put in excess of about 40 hellfire missiles on the target, a couple hundred Flechette rockets and a lot of 30 millimeter and then broken through the low level back. Destroying the radar towers opens up a (14) _____ of the sky 20 miles wide. About 5 minutes into the flight back, we could see the jets coming right over our head. And that's how the air war started. Over 900 coalition aircraft make their way into Baghdad Safe (15) _____ courtesy of the Apache.

Task 6   Draw a mind map to describe the attack process of Army Aviation at the very beginning of *Operation Desert Storm*.

Task 7   Talk and Present.

Scenario 1

Suppose your team members all work as staff officers in the Operation Room of an Army Aviation Brigade, your task is to take out an enemy's air-defense battery and your team members are engaged in a heated discussion to make plans of the attack before reporting it to the commander. Make a conversation based on the words and sentences you have learned in the previous reading and listening materials.

Scenario 2

Please find another example to illustrate Army Aviation attack by employing Manned-Unmanned Teaming (MUM-T) or a single armed UAS.

Delta Air Assault

Scan for listening resources

An air assault is the movement of friendly assault forces by Rotary Wing (RW) aircraft to engage and destroy enemy forces or to seize and hold key terrain. Air assaults extend the tactical and operational reach of the combined arms team by overcoming the effects of terrain, achieving surprise, and isolating, dislocating, or destroying enemy forces by rapidly massing combat power at the maneuver commander's time and place of choice.

Army Aviation conducts air assaults in support of offensive, defensive, and stability operations throughout the depth and breadth of the AO. Air assaults are combined arms operations conducted to gain a positional advantage, envelop, or turn enemy forces that may or may not be in a position to oppose the operation. Assault units may be tasked with air assaulting a Tactical Combat Force (TCF) in order to counter Level III enemy penetrations of the main battle area.

Task organization of the aviation task force supporting the Army Aviation Task Force (AATF) is based on mission variables but at a minimum always includes an assault element and an attack or reconnaissance element as the foundational aviation maneuver capability.

The assault element may be made up of assault helicopters, heavy lift helicopters, or a combination of both. Aviation assault and heavy lift units transport ground maneuver forces and equipment from secure or permissive PZs to either unsecure or secure LZs in the objective area. Based on mission variables and the AATF commander's intent, LZs may be directly on or very near the objective or offset from the objective. The closer the LZ is to the objective, the greater the ability to rapidly mass combat power and with greater likelihood of achieving surprise. Offset LZs are chosen when no suitable LZs are available, to enhance survivability during the landing phase if the threat on the objective is high or when the AATF commander desires to infiltrate into the objective. However, significant offset distances between the objective and LZ location may reduce the element of surprise, may require a larger ground tactical force, and may allow the enemy early warning and freedom to maneuver to gain a position of advantage. Availability, size, and suitability of LZs; size, disposition, and capabilities of the enemy; size and capabilities of the AATF; and the AATF commander's intent drive the determination of LZ locations.

Task 1　Read and answer true (T) or false (F).

1. Basically, an air assault is a movement of assault forces to a certain area to conduct an offensive operation. _____

2. Comparatively, an air assault is a much quicker reaction forces than the mechanized infantry. _____

3. Air assaults are joint operations conducted to gain a positional advantage or to impose disadvantages on the enemy. _____

4. The assault element may be made up of assault helicopters, heavy lift helicopters, or a combination of both. _____

5. Offset LZs are chosen when the AATF commander desires to infiltrate into the objective. _____

🎧 4-D-1

Task 2　Listen and answer the questions.

1. For how long has Bin Laden been wanted?
2. Where did Bin Laden live before US Navy SEALs stormed his hideout?
3. What was the purpose of the Operation "Geronimo"?
4. When did the US Navy SEALs started the operation?
5. Why the Pakistan government wasn't told about the raid?

🎧 4-D-2

Task 3　Listen and answer true (T) or false (F).

1. The raid was delayed one day because of bad weather. _____

2. The raid was conducted from Afghanistan with the target in Pakistan. _____

3. The black hawks took off with escorting Apaches in radio silence. _____

4. The Army pilots were simply guided by night vision goggles. _____

Task 4   Listen again and answer the questions.

1. When was the raid conducted?

2. Who planned and commanded this operation?

3. Where did they conduct the raid?

4. How did the pilots fly the helicopter?

🎧 4-D-3

Task 5   Listen and answer true（T）or false（F）.

1. The raid conducted on that night is no different from others in Afghanistan and Iraq. _____

2. The US asked the Pakistani troops to cooperate with the raid. _____

3. The team conducting the raid were 24 seals and a dog. _____

4. The assault element is made up of 2 modified Black Hawk. _____

Task 6   Listen again and complete.

Chalk One, which is the one I was on, was gonna hover over the (1) _____ here. We will drop the two fast (2) _____ here, (3) _____ down the ropes into the courtyards here and then go about our businesses while Chalk Two would (4) _____ out here, just over here by the road, drop the (5) _____ containment team off. They will provide security external. We'd have two men and our combat assault dog who will do a quick (6) _____ of the perimeter down to the south and around to make sure there were no (7) _____ underneath the walls in case somebody dig it here for them to have time to escape. After dropping those guys off, the second helo will come up and hover over the third floor, drop off the remaining guys. They would (8) _____ right down into the balcony, assaulting from top-down and our guys assault from the bottom up.

Task 7   Talk and Present.

Please compare the roles of Army Aviation in *Operation Desert Storm* and *Operation Neptune Spear* and summarize the differences between Attack and Air Assault.

Echo Reconnaissance

Reconnaissance is a mission to obtain, by visual observation or other detection methods, information about the activities and resources of an enemy or adversary, or to secure data concerning the meteorological, hydrographic, or geographic characteristics of a particular area. Reconnaissance missions enable the commander to gain comprehension of the situation and form a mental image of the battlefield by addressing crucial gaps in information, reducing potential risks, allocating resources effectively, and determining task priorities. Effective reconnaissance enables the commander to identify where the enemy is weak or strong, the best place or opportunity to concentrate combat power to gain and maintain a position of relative advantage, or where and when to best deny the enemy a position of relative advantage.

Army Aviation conducts reconnaissance as part of its parent organization's focused information collection efforts by either fighting for or collecting information by stealth and observation. Reconnaissance is conducted before, during, and after operations to assist the commander with the formulation, confirmation, or modification of a course of action (COA).

Army Aviation reconnaissance units are specifically equipped, trained and organized to conduct all forms of reconnaissance except special reconnaissance. Irrespective of whether reconnaissance is officially assigned or not, it is always an implicit duty for every aviation unit to collect and relay information about enemy and friendly positions, terrain, and civilian activities witnessed throughout all operations.

Army Aviation conducts reconnaissance at all echelons either independently as a pure aviation maneuver force or as part of a deliberately planned scheme of maneuver as a member of the combined arms team. The size of the aviation reconnaissance force is driven by the size of the AO, complexity and number of reconnaissance objectives, fidelity of the information required by the commander, the enemy situation, and the time available to answer the commander's specified information requirements.

Commanders may seek additional information regarding a terrain feature, geographic area, or enemy force as part of their reconnaissance objectives. Based on the capabilities of the reconnaissance force and time available to conduct the reconnaissance, the commander and staff further delineate the priority of tasks and information collection efforts to ensure the most critical information is collected to enable timely decisions.

Task 1 Read and answer the questions.

1. What are the general purposes of conducting reconnaissance missions?

2. What kind of information do commanders normally seek out through reconnaissance?

3. To what degree does Army Aviation get involved in reconnaissance missions?

4. Are Army Aviation reconnaissance units organized to conduct all forms of reconnaissance?

5. What elements may affect the size of the aviation reconnaissance force?

6. What are the common reconnaissance objectives?

7. How do Army Aviation commanders prioritize their information collection efforts?

 4-E-1

Task 2 Listen and choose the best answer.

(   )1. This passage introduces _____ types of reconnaissance.

    A. three B. four C. five

(   )2. The form of reconnaissance that focuses on obtaining detailed information about the terrain or enemy activity within an assigned area is called

    _____.

    A. zone reconnaissance

    B. area reconnaissance

    C. route reconnaissance

(   )3. The form of reconnaissance that involves a directed effort to obtain detailed information on all routes, obstacles, terrain, and enemy forces in an area defined by boundaries is called _____.

    A. zone reconnaissance

B. area reconnaissance

C. route reconnaissance

(　　)4. Area reconnaissance is normally _____ than zone reconnaissance.

A. less complex

B. more sophisticated

C. much larger

Task 3　Listen again and answer true（T）or false（F）.

1. Commanders assign route reconnaissance missions to gain detailed situational understanding when the enemy situation is vague or the understanding of the terrain is limited. _____

2. The major difference between an area and zone reconnaissance is that in a zone reconnaissance the unit has to first move to the place，then conduct the reconnaissance. _____

3. The route reconnaissance provides information about any friendly，enemy，or civilian activity along the route. _____

4. Route reconnaissance is conducted as a combined arms operation at least at the battalion task force. _____

5. Compared with area reconnaissance，reconnaissance in force focuses more on the enemy forces but is less likely to get involved in any form of engagement. _____

🎧 4-E-2

Task 4　Listen to the description of a reconnaissance mission and complete the table with the information you've heard.

Time	
Weather	
Visibility	
Terrain	
Situation	
Mission	

4-E-3

Task 5　Watch the execution of the reconnaissance mission and try to identify what has been done during different phases of the mission.

> A. describes the target and unidentified tank
>
> B. fires the cannon
>
> C. enters its UTM coordinates into the system's computer
>
> D. locates the tank and uses the laser, then activates the data link
>
> E. uses natural terrain to avoid enemy fire
>
> F. locks his sight and slaves cannon out of the target for automatic tracking
>
> G. notes and reports river traffic and movement to the north, and marks them on the map block
>
> H. selects one of the checkpoints near the assigned area
>
> I. switches back to the night vision system, illuminating the area with covert light

1. Phase Ⅰ: After boarding the aircraft, the pilot (1) _____.

2. Phase Ⅱ: When rising to altitude for a search with radar and moving target indicator, the observer (2) _____.

3. Phase Ⅲ: When descending for closer examination, the pilot (3) _____ and the observer (4) _____.

4. Phase Ⅳ: When proceeding north from the river to further identify objects in hot spots, the observer (5) _____.

5. Phase Ⅴ: After identifying the objects as a tank, the observer (6) _____.

Task 6　Listen again and answer the following questions.

1. What forms of reconnaissance mission do the pilot and the observer execute?

2. What is the reconnaissance aircraft equipped with?

3. What does the observer note and report when the reconnaissance aircraft is rising to altitude for a search?

4. What do the pilot and the observer do after identifying the river traffic?

5. How does the observer deal with the tank he identified?

4-E-4

Task 7 Listen and fill in blanks with the exact words you've heard.

While rising to altitude for a search with radar and moving target indicator, the observer (1) _____ and (2) _____ river traffic and movement to the north. He (3) _____ these on his map block. They then (4) _____ for closer examination. Using the night vision system in the magnifying periscope, the river traffic is identified. While the pilot uses natural terrain to avoid enemy fire, the observer, utilizing the laser range finder, (5) _____ his sight and slaves cannon out of the target for automatic tracking, then he (6) _____ the cannon.

Proceeding north from the river, the observer uses his forward-looking infrared sensor. He chose several hot spots. To further identify these objects, he (7) _____ back to the night vision system, (8) _____ the area with covert light.

They are identified as moving vehicles, so the observer zooms in on the target to see a tank. From standoff distance, the observer (9) _____ the tank and uses the laser, then (10) _____ the data link. Automatically the position of the target is (11) _____ to the intelligence center. By voice radio, the observer (12) _____ the target and unidentified tank. A killer aircraft is assigned, meanwhile intelligence requests photo identification of the tank, if possible.

Task 8 Talk and present.

Suppose you are the observer from the above reconnaissance mission. After identifying the moving tank, you locate the tank and report the situation to the intelligence center. Then you are told a killer aircraft will be assigned and the intelligence center requests for photo identification of the tank. Now make up a dialogue between the observer and the intelligence center with the prowords and sentence patterns you've learned in this lesson.

Foxtrot Movement to Contact

Scan for listening resources

An offensive task, a movement to contact (MTC) is designed to develop the situation and establish or regain contact. A movement to contact is executed when the commander finds it difficult to decide whether an attack can be launched or not since the enemy situation remains vague or not specific. Therefore, a movement of contact helps prevent the premature commitment of friendly combat power.

Conducting a movement to contact enables freedom of action to develop the situation and creates favorable conditions to carry out subsequent tactical or enabling tasks. The speed, range, lethality, long-range communications, and persistent reconnaissance capabilities of Army Aviation attack or reconnaissance units, using MUM-T, make them ideally suited to conduct movement to contact. Army Aviation executes movement to contact at the platoon to battalion or squadron level, either independently, or as a member of the combined arms team. Regardless of the scenario, the approach towards making contact is structured in a manner where the minimum required security force is deployed ahead to establish and sustain initial contact with the enemy and a separate force capable of developing the situation based on the size of the expected enemy force and commander's intent.

Movement to contact mission requires the commander not to have contact with the enemy main body. However, the commander may still know the location of at least some enemy reserve and follow-on forces. If the corps or division commander has enough intelligence information to target enemy uncommitted forces, reserves, or sustaining operations activities, the commander normally designates forces, such as long-range artillery systems, attack helicopters, extended range unmanned aircraft, and fixed-wing aircraft to engage known enemy elements regardless of their geographical location within the AO.

The movement to contact is often planned and executed more like a zone reconnaissance. Different from a zone reconnaissance, the effort centers on locating the enemy force, developing the situation early, and preventing the premature deployment of the brigade combat teams main body. Terrain reconnaissance is conducted as necessary to support finding the enemy. As a result, movement to contact proceeds much faster than a zone reconnaissance.

Task 1   Read and answer the questions.

1. What is the general purpose of conducting a movement to contact?

2. In what circumstances do commanders deploy a movement to contact?

3. Why are Army Aviation units capable of conducting a movement to contact?

4. Are the units supposed to have contact with the enemy main body in movement to contact mission?

5. What is the typical size of the army aviation unit executing a movement to contact?

6. Please compare a movement to contact with a zone reconnaissance.

 4-F-1

Task 2   Listen and take notes.

Time: June. 11, 1430 hrs-1630 hrs
Lecturer: (1) _____ (PMS3 instructor)
Topic: (2) _____

Ⅰ. (3) _____ **of this lecture**
(4) _____ a doctrinally sound way to (5) _____ at the (6) _____ level.

Ⅱ. (7) _____ **to use in this lecture**
① ATP 3-21.8 (8) _____;
② the ranger handbook TC 3-21.76.

Ⅲ. **Importance of movement to contact**
① one of the five (9) _____ tasks we can conduct in the (10) _____ level;

② one of the (11) _____ tasks we can conduct at an (12) _____
camp.

IV. Purpose of movement to contact

Since we may not know the exact (13) _____ or the exact (14) _____
of the enemy, we're trying to (15) _____ or (16) _____ contact
typically in order to (17) _____ the enemy.

V. Types of movement to contact

① (18) _____ and (19) _____ method;
② cordon and search method.

Focus of this lecture: (20) _____.

 4-F-2

Task 3 Listen and complete the fundamentals of conducting movement to contact
with the information you've heard.

Item	Fundamentals of conducting movements to contact
1	Focus all efforts on (1) _____.
2	Gain enemy contact with the (2) _____ within the (3) _____ time
3	Avoid (4) _____ of the main body on ground chosen by the enemy.
4	Destroy, defeat, (5) _____, (6) _____, or delay enemy forces within capability or in accordance with the (7) _____.
5	Keep (8) _____ forces within supporting distances to (9) _____ a flexible response.

4-F-3

Task 4 Listen and answer true (T) or false (F).

1. The speaker in this passage introduces search and attack as a specific

technique for attack and reconnaissance mission. _____

2. The search and attack technique is especially effective when the enemy is in large teams over a limited area. _____

3. The essence of the search and attack is to detect, disrupt, or destroy the enemy. _____

4. The key for reconnaissance units to locate the enemy and find its composition is to do their job covertly. _____

5. To halt the enemy force's movement along its most usual route of departure is a practical tactic in fixing the enemy. _____

6. To better destroy the enemy, attack and reconnaissance units are supposed to focus merely on ensuring their combat power and capabilities. _____

4-F-4

Task 5　Listen and complete.

A Blue 14, this is Blue 6. (1) _____ . Over.

B Blue 14. (2) _____ . Over.

A (3) _____ . Message. Three (4) _____ and multiple (5) _____ are (6) _____ TRDP 016. Request further instructions. Over.

B Blue 14. (7) _____ . Can you identify any civilians around the area? Over.

A Blue 6. (8) _____ . Over.

B Blue 14. At my command, fire. (9) _____ .

A Blue 6. (10) _____ . Over.

B Blue 14. Roger. Out.

Task 6　Talk and present.

As an instructor in an advanced camp, you are going to give cadets a lecture on movement to contact. Your lecture should involve the following information:

1. Purpose of movement to contact

2. Principle of conducting movement to contact

3. Procedures of conducting search and attack (one of the common methods in movement to contact)

Golf Security

Scan for listening resources

Security tasks refer to tasks performed by the commander to warn about enemy operations early and accurately. These tasks give the protected forces enough time and maneuver space to react to the enemy. Security tasks also help the commander understand the situation better so they can use their protected force effectively. These tasks are meant to prevent surprises and reduce uncertainties. The protected force includes friendly ground maneuver forces, facilities and the local people.

Security tasks can be divided into five types: screen, guard, cover, area security, and local security. Army Aviation attack or reconnaissance units are specially equipped, trained, and organized to carry out security operations. However, aviation units can only independently perform the screen task.

Screen is a security task to give the protected force an early warning if the enemy is coming. It's used when there's less probability of making contact with the enemy, when the enemy force is expected to be small, or when the protected force needs only a short time to react. Screens can be either stationary or on the move. Stationary screens are conducted to the front, flanks, or rear of a protected force that's not moving or to the flanks or rear of a moving protected force. A moving screen is linked to how fast and in what direction the protected force is moving, and is conducted either to the rear or to the flanks of the moving force.

Guard is a security task to protect the main body by fighting to gain time, while also observing and reporting information about the enemy. It also aims to stop the enemy from observing or shooting at the main body. When Army Aviation is supporting or doing a guard task, they do a wide range of tactical, supporting, and sustaining tasks. These include attacking, scouting, making contact with the enemy, conducting a screen, air assaults, air movement, aerial C2, and AE or aerial CASEVAC.

Cover is a security task to protect the main body by fighting to gain time, while also observing and reporting information about the enemy. It also aims to stop the enemy from observing or shooting at the main body. Aviation forces can be organized to support a larger military unit tasked with a cover mission. When Army Aviation supports a cover mission, they carry out a full range of aviation tasks to assist the combined arms team, including attacking, scouting, making contact with the enemy, conducting a screen, air assaults, air movement, aerial C2, and AE or aerial CASEVAC.

When planning and carrying out security operations, using the following five principles to guide the process and help successfully complete security missions: give an early and precise warning; allow time to react and room to move; focus on the protected force, area or facility; keep scouting and gathering information constantly and maintain enemy contact.

Task 1   Read and answer true (T) or false (F).

1. Basically, security tasks are those tasks performed by the commander to provide early and accurate warning of enemy operations. _____

2. The protected force ranges from friendly ground maneuver forces and facilities to the local population. _____

3. Security consists of four tasks: screen, guard, cover and area security. _____

4. Aviation units can only independently perform the guard task. _____

5. Both guard and cover are security tasks that involve protecting the main group by fighting to gain time. _____

Task 2   Read and answer the questions.

1. When will the screen task be used?

2. What are the types of the screen task?

3. What are the similar operations shared by the guard task and the cover task?

4. What are the principles of planning and carrying out security operations?

 4-G-1

Task 3   Listen and answer true (T) or false (F).

1. A screen is utilized when the likelihood of enemy contact is minimal. _____

2. A screen is only used to cover gaps between forces. _____

3. One of the critical tasks for an aviation security force conducting screens is to prevent any enemy from passing through undetected and unreported. _____

4. The Army Aviation units are not suited to operate as an independent screening force. _____

Task 4　Listen again and complete.

A screen is used to cover gaps between forces, exposed flanks of (1) _____ forces, or to the rear or flanks of a moving force. Screens are used when the expectation of enemy contact is low, the enemy force is expected to be small or the protected force requires minimal (2) _____ time. The enhanced endurance, (3) _____, lethality, sensors of Army Aviation attack or (4) _____ units make them ideally suited to operate as an independent screening force or as a part of a (5) _____ arms team conducting security operations.

Critical tasks for an aviation security force conducting screens, include the following: ① Allow no enemy to pass through the screen (6) _____ and unreported; ② Maintain continuous (7) _____ of all avenues of approach larger than a designated size into the security area; ③ Destroy or (8) _____ all enemy reconnaissance prior to the enemy gaining observation on the protected force.

🎧 4-G-2

Task 5　Watch and answer true (T) or false (F).

1. Uniform 34 is approaching staging area. _____

2. Both 64 and 65 are going into holding pattern. _____

3. Both 61 and 62 take up overhead pattern to provide sniper cover. _____

4. One bird is hit by the RPG. _____

Task 6　Watch again and answer the following questions.

1. When do the little birds touch down at target?

2. What is the mission of the little birds?

3. Why does a soldier fall down from the bird?

4-G-3

Task 7　Listen and complete.

　　Based on the protected force commander's (1) _____ reaction time, Army Aviation operates at (2) _____ distances from the main body thus (3) _____ additional time and space for the protected force commander to make an (4) _____ decision to employ forces. Based on the commander's (5) _____, the aviation security force may (6) _____ to conduct offensive tasks to (7) _____, delay, or disrupt the enemy forces' tempo and cohesion, providing reaction time and (8) _____ space to the protected force.

Task 8　Talk and present.

　　Suppose you are a team leader who will brief your Army Aviation troops on a security mission. Your task is to transport, cover and evacuate the SEAL teams to arrest a group of hostile elements. Make a briefing based on the knowledge, words and sentences you have learned according to the briefing format.

Hotel Air Movement

Scan for listening resources

An air movement is the air transport of units, personnel, supplies, and equipment including airdrops and air landings and it is different from air assault. Air movement operations are a feasible means of transport and distribution, which can provide support for offensive, defensive, and stability. Loads can be configured internally or externally. It depends on different missions and types of aircraft to conduct the air movement operation. Army Aviation rotary-wing aircraft conduct air movement using both internal and external loads. Army Aviation fixed-wing aircraft conduct air movement with internal loads to move limited important personnel and supplies. The aviation unit finally determines whether the load is carried or not and also decides in advance what portion of the load is to be carried internally or externally.

Air movement operations can be conducted in support of a variety of operations, such as foreign humanitarian assistance, foreign disaster relief, homeland defense, non-combatant evacuation, battlefield circulation of key leaders. Air movement operations allow the ground force commander to control the pace of operations and tackle the enemy force at the established time and place. Utility and cargo helicopters are used to supplement ground transportation and support with offensive and defensive operations. Therefore, the supported commander can overcome difficult terrain and time limitations on operations.

A typical air movement may be easily attacked by enemy forces as most of missions are to support troops and transport equipment to secure areas. Air movements are not as complex as an air assault operation in planning and execution, but a detailed plan is also needed to ensure the safety of the crew and passengers. The size and condition of the pickup zone (PZ) and security of the landing zone (LZ) must be considered in the planning process. Air movement also requires coordination in advance between the operations elements of aviation units and the units supported.

Task 1　**Match the words or phrases with the definitions.**

| airdrop | transport helicopter | airland | relief goods |

1. A term used in military operations to describe the process of deploying and transporting troops, equipment, and supplies via aircraft to support ground-based mission. _____

2. The act of dropping supplies, soldiers, etc. from an aircraft by parachute. _____

3. A kind of helicopters designed to move cargo over large distances, including oceans. _____

4. Food, water, medication, clothes, etc. given to people in need, especially in disaster areas. _____

 4-H-1

Task 2　**Listen and choose the best answer.**

(　)1. When were the relief goods transferred to the place where PAAC regiment was stationed?

A. Before dawn on May 13.

B. On the morning of May 13.

C. On the morning of May 14.

D. On the late evening of May 14.

(　)2. What was the difficulty PAAC regiment faced with when they arrived in the destination?

A. Air dropping the materials.

B. So many victims of the disaster.

C. Complex terrain.

D. So many tents on the ground.

(　)3. What decision did the crews make when facing with the difficulty?

A. To deliver the goods later.

B. To airland on an open terrace.

C. To take off for the return flight.

D. To soar into the sky.

( )4. How many people were transported in the earthquake relief?

    A. More than 50,000 people.

    B. More than 5,000 people.

    C. More than 15,000 people.

    D. More than 150,000 people.

Work in pairs. Discuss about the role and status of PAAC in disaster relief.

 4-H-2

Task 3   **Listen and answer the questions.**

    1. What request do Captain Brian Lutz and his crew receive?

    2. What is "speed balls"?

    3. What makes CH-47 exposed and vulnerable?

Task 4   **Listen again and complete.**

    Sharana Air Base Afghanistan. Captain Brian Lutz and his crew ready their CH-47 for a (1) _____. Thirty miles to the south, a platoon of infantry soldiers (2) _____ a battle with the enemy on top of a mountain (3) _____ 8,000 feet high and they're almost (4) _____. Lutz and his crew quickly load their Chinook with cargo they call "speed balls". (5) _____ and loaded, the Chinook lifts off and (6) _____ towards the fire fight. Every minute (7) _____ for the men on the ground. Lutz and his crew fly in the dangerous territory, but there's a problem. Chinook's size makes it a very (8) _____ target. The Chinook's (9) _____ make it the go-to bird for the most dangerous operations. But as a result some of the largest (10) _____ of life in Afghanistan have been in Chinooks.

Task 5   **Work in pairs. Discuss about advantages and disadvantages of Chinooks in air movement.**

Task 6   **Talk and present.**

    Work in pairs. Create a radio communication script based on the operation in Task 4 and then perform the dialogue.

India Aeromedical Evacuation

Scan for listening resources

Medical evacuation (MEDEVAC) is the process of moving patients while providing them en-route care, which is a timely and efficient movement by medical personnel. The provision of en-route care can greatly enhance the patient's possibility of survival and recovery. The Army MEDEVAC system consists of standardized MEDEVAC ambulances (ground and air). These ambulances have been designed, staffed, and equipped to provide en-route medical care and are used exclusively to support medical missions.

Army air ambulance units provide direct support and area support, which are tasked to locate, acquire, treat, and evacuate patients to appropriate areas. Those who use Army air ambulances are classified as non-combatants, which can reduce the risk of the patient and crew when conducting missions. The air ambulance unit operates in an alert status to rapidly respond to evacuation missions and is not diverted to perform any other tasks. Aeromedical evacuation (AE) is generally preferred for seriously wounded, injured, and ill personnel because of its speed, range and flexibility.

Different from MEDEVAC, aerial casualty evacuation (CASEVAC) is the unregulated transport of injured personnel by using Army Aviation assets that do not have onboard medical personnel or equipment. CASEVAC is a part of combat health support. CASEVAC includes battlefield pickup of casualties, evacuation of casualties to initial treatment facilities and subsequent movement of casualties to treatment facilities within the combat zone. CASEVAC is an aviation mission directly supporting a ground unit with casualty evacuation aircraft from forward locations to the brigade support area (BSA) or other designated treatment facility. CASEVAC can be conducted by any Army Aviation utility aircraft when tasked by the commander. It is advisable that the least severely injured are evacuated using CASEVAC assets.

Task 1　Read and choose the best answer.

(　　)1. What is the purpose of medical evacuation (MEDEVAC)?

　　　A. To provide en-route care and timely movement of patients.

B. To transport patients without any medical personnel or equipment.

C. To support medical missions with ground and air ambulances.

D. To evacuate patients to appropriate areas.

(　　)2. How are Army air ambulance units classified?

　　A. Non-combatants.

　　B. Combatants.

　　C. En-route care providers.

　　D. Ground support units.

(　　)3. What is the main advantage of aeromedical evacuation (AE)?

　　A. Its ability to provide en-route medical care.

　　B. Its speed, range and flexibility.

　　C. Its use of standardized ambulances.

　　D. Its support for combat health missions.

 4-I-1

Task 2　Watch and answer true (T) or false (F).

1. The ability to evacuate and care for the wounded on the battlefield has long been a top priority. _____

2. Major Carter Harman was the first pilot who completed a modern medical evacuation mission. _____

3. During the Korean War, the casualty death rate was 2.5 deaths to every one hundred casualties. _____

4. During World War II, medics had enough room inside the aircraft to treat the patients. _____

5. The HH-60 Medevac Blackhawk helicopter is equipped with onboard life support systems. _____

Task 3　Watch and fill in the table with the correct information.

Period	Details about MEDEVAC	Significance
World War II (1944)	The first modern medical evacuation (1) _____ carried out by Lieutenant Carter Harman.	The (2) _____ of helicopter medical evacuations.

Period	Details about MEDEVAC	Significance
Korean War (1950–1953)	The first (3) _____ use of helicopters for medical evacuation during the war in Korea, where an estimated (4) _____ casualties were evacuated.	A significant (5) _____ of the casualty death rate from 4.5% in WWII to 2.5%.
Vietnam War	The (6) _____ deployment of Huey medical evacuation helicopters during the war, allowing (7) _____ to continue treating patients on the way to field hospitals.	Another reduction in (8) _____ rate.
Today	The advanced (9) _____ life support systems of the HH-60 Medevac Blackhawk helicopter.	The continued (10) _____ and advancement in medical evacuation capabilities.
In the future	DARPA's Wound Stasis Program developing a biological (11) _____ to control severe internal bleeding caused by IEDs.	The potential to save (12) _____ lives.

Task 4   Listen again and complete.

1. The first medical evacuation by _____ occurred during World War II in the Pacific _____.

2. The first _____ use of helicopters for medical evacuation took place during the war in _____.

3. The mass _____ of the Bell UH-1 Huey helicopters led to another reduction in _____ rate.

4. Today the speed at which the wounded are _____ on the battlefield

continues to _____.

5. DARPA is developing a biological _____ to control internal _____.

🎧 4-I-2

Task 5   Listen and answer the following questions.

1. What preparations do the crew members do before launching the aircraft for a mission?

2. How long does it typically take for the crew to launch the aircraft for a mission?

3. What is the main goal during the flight to the hospital?

Task 6   Listen again and answer true (T) or false (F).

1. The medevac team receives advance notice before someone gets hurt.

_____

2. The medevac team takes about 20 seconds to reach a speed of 150 miles per hour. _____

3. The medevac team always flies at night. _____

4. Once on the ground, it takes the medevac team 3-5 minutes to load the patient onto the aircraft. _____

5. The medevac team only provides medical interventions if necessary.

_____

🎧 4-I-3 & 4-I-4

Task 7   Listen and complete the following two conversations.

Conversation 1

A We need to carve out an (1) _____ right over there.

B Sir.

A Blow the trees down. (2) _____ in the hole.

C Colonel, this is Snakeshit and "too tall". We're coming in with (3) _____.

A Crandall, we've (4) _____ a new LZ. When you come in, come in from the east. Out.

Conversation 2

A Come on! We'll give you a hand! We're moving (5) _____.

B All right, get that ammo off and get the (6) _____ on. Bring them on! Get them on, boys.

A No, we're (7) _____!

C Leave him! I'll get out. It's Ray. He's hurt worse than me. You (8) _____! I'll see you back there, Ray.

Task 8   Listen again and use no more than three words to describe the missions conducted by aviation in two conversations.

1. _____

2. _____

Task 9   Talk and present.

Suppose you are required to conduct a casualty evacuation task. Make a conversation in such a scene within your group with the prowords and sentence patterns in two conversations.

Juliet Personnel Recovery

Army personnel recovery (PR) is the military efforts taken to prepare for and execute the recovery and reintegration of isolated personnel. It includes five tasks — report, locate, support, recover and reintegrate. Army Aviation plays a role in executing established procedures to accomplish these tasks. There are four methods of recovery used by Aviation forces to support the ground force commander or to recover their own personnel.

Unassisted recovery comprises actions taken by isolated personnel to achieve their own recovery (sometimes referred to as self-recovery). Isolated personnel independently evacuate to return to friendly forces or to a location where they can successfully connect with friendly forces or be recovered through another method.

Immediate recovery refers to the actions taken to locate and recover isolated personnel by forces directly observing the isolating event or, through the reporting process, determining isolated personnel are close enough for them to conduct a rapid recovery with the forces at hand without detailed planning or coordination.

Deliberate recovery is conducted by Army forces when immediate recovery is unfeasible or unsuccessful. A deliberate recovery needs to be carefully planned and coordinated due to various considerations, such as weather, enemy activities, isolated personnel situation, current operations, and the capabilities of the recovery force.

External supported recovery is conducted when immediate or deliberate recovery is unfeasible or unsuccessful. It is either the support provided by the Army to other joint task force components, interagency organizations, or multinational forces, or the support provided by these entities to the Army.

All echelons above battalion have trained PR specialists who are assigned as PR officers. Commanders at all subordinate echelons assign a PR representative. The PR representative serves as the unit's PR program manager to ensure all PR tasks are planned, coordinated, and completed.

Task 1 Read and answer true (T) or false (F).

1. Army personnel recovery includes five tasks: report, locate, support, recover, and reintegrate. _____

2. Unassisted recovery is when isolated personnel rely on their own actions to achieve recovery. _____

3. Immediate recovery requires detailed planning and coordination. _____

4. Deliberate recovery is conducted when immediate recovery is feasible and successful. _____

5. External supported recovery comprises the support provided by the Army to other joint task force components. _____

🎧 4-J-1

Task 2 Discuss the importance of PR and then listen and complete.

"Personnel recovery is a (1) _____. It (2) _____ of those fighting on the ground and in the air to (3) _____, knowing that someone will always (4) _____. Critical to the warfighter is knowing that a (5) _____ PR force is (6) _____ at a moment's notice, willingly (7) _____ ... so that others may live."

🎧 4-J-2

Task 3 Listen and complete the conversation.

A Hello C20. This is C21. (1) _____. Over.

B C20. Send. Over.

A C21. (2) _____. Co-pilot (3) _____ at grid 987624. Near small house. We request urgent personnel recovery. Over.

B C20. Question. Is the area safe? Over.

A C21. (4) _____. There are some (5) _____ in the vicinity and two of us are almost (6) _____. Over.

B C20. We are sending a fire support team and an (7) _____ team. Seek shelter. Acknowledge. Over.

A C21. Wilco. Out.

Task 4   Discuss in groups. Suppose a helicopter is shot down by the enemy and as the PR representative, what relevant information requirements should you identify before the PR operation?

Sample of a Downed Aircraft Recovery Report

Call sign of sender			
Cause of aircraft incident (If known)			
Isolated personnel location (grid or distance from known landmark)			
Sierra (secured) or November (not secured)			
Personnel number			
condition	A-Wounded in Action	B-Killed in Action	C-Missing in Action
Aircraft	A-Communications security (COMSEC) status destroyed/ recovered/zeroized	B-Accessibility by vehicle/by air (closest LZ)	C-Damage assessment recoverable/not recoverable/fire
Terrain			
Threat situation at site (Known or suspected enemy activity)			

Task 5   Make a conversation about reporting and locating in a crash with the help of the map and the sample above.

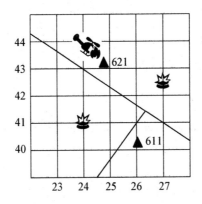

Task 6   **Watch and tell the method of PR. Then complete the conversation.**

A 25，this is 64. Over.

B 25. Over.

A 25，Chalk Four is (1) _____. Over.

B 64，I can't see the crash site. Over.

A 25，(2) _____. You will find it. (3) _____. Take the rest of your chalk to the crash. Check for survivors. (4) _____. All other chalks will follow. 25，(5) _____? Over.

Task 7   **Put the following sentences in order.**

☐ The cargo helicopters land and the special forces disembark.

☐ The attack helicopters reach the area，identify the enemy positions located nearby and attack them to clear the way to the rest of the rescue team.

☐ When the ground team signals they are ready for the exfiltration，the cargo helicopters land again and leave with attack helicopters escort.

☐ Once the area is deemed safe from enemy forces，the reconnaissance is carried out to locate the isolated personnel.

☐ The cargo helicopters take off again and start patrolling the area while the team on the ground identifies the isolated personnel and provides medical support，if needed.

☐ In the meanwhile，the cargo helicopters with the extraction forces have reached the area.

☐ The attack helicopters provide armed overwatch as the rescue mission progresses. They are ready to engage the enemy if needed.

Task 8   **Talk and present.**

Suppose you are the PR representative. Please give a mission briefing on the plan of recovering the personnel in a crash. The following sample is for your reference.

> **Personnel recovery team mission briefing**
> • *Team roll call*
> • *Location of isolated personnel and downed aircraft*

- *Map briefing of surrounding area*
- *Develop a route if deployment by ground*
- *Enemy situation*
- *Hazards (mines, weather, traffic)*
- *Friendly situation*
- *Method of recovery (unassisted / immediate / dedicate / external recovery)*
- *Convoy briefing and request*
- *ETD and ETA*
- *Weather forecast for next 24 to 48 hours*
- *Communications frequencies and call signs*

Kilo Command and Control Support

Scan for listening resources

Army Aviation Command and Control (C2) assets enable C2 to be comprehensive and offer voice and data communications that go beyond line of sight. Army Aviation enhances a commander's capacity to integrate and synchronize operations by providing C2 aircraft, unmanned aircraft system (UAS), communication relay package and air traffic services (ATS).

Maneuver commanders can more effectively understand, visualize, describe, lead, and assess operations over extended ranges and complex terrain with the use of command, control, and communications aircraft. The ability for air and ground commanders to rapidly traverse and observe the area of operations (AO) is provided by the command aviation company (CAC). In order to accomplish this, the CAC provides aerial retransmission and airborne C2 as directed. The C2 UH-60 aircraft gives the commander an enhanced capability to communicate over extended distances by performing C2 while moving, serving as an aerial tactical command post (CP), and providing an early entry CP. The onboard communications linkages allow for continuous contact between the commander and committed forces. Communication systems enabled by networks allow commanders and staffs to absorb much more information more quickly and clearly.

The aviation unit providing the aircraft must coordinate with the supported unit early to integrate the C2 aircraft during the planning process. The aircrew of the C2 aircraft should attend orders and rehearsals of the supported unit to fully understand the operational scheme of maneuver and to best integrate the aircraft into the plan.

Task 1　Read and answer true (T) or false (F).

1. Army Aviation Command and Control assets can provide beyond line of site voice and data communications. _____

2. Army Aviation provides only unmanned aircraft systems (UAS). _____

3. Command, control, and communications aircraft help maneuver commanders understand and assess operations over extended ranges and complex terrain. _____

4. The onboard communications linkages in the C2 UH-60 aircraft allow for continuous contact between the commander and committed forces. _____

5. The C2 aircraft is integrated into the plan without any coordination. _____

Task 2   Work in pairs and discuss why command and control support is so important.

4-K-1

Task 3   Watch and work in groups to draw a mind map based on the general's mission briefing. Then try to give a mission briefing based on your mind map.

4-K-2

Task 4   Watch and answer the questions.

1. What is the action aimed for?

2. What is the role of Colonel Harrell in this action? How does he accomplish that?

4-K-3

Task 5   Watch and choose the best answer.

(   )1. What is the call sign of the helicopter in the video?

      A. Super 64.   B. Wolcott.

      C. Star 41.   D. Juliet 25.

(   )2. Why does the helicopter need to be checked out on the airfield?

      A. It got clipped pretty good.

      B. It can't fly.

      C. The instrument panels are okay.

      D. The tail rotor is gone.

(   )3. What happens to the helicopter at the end?

      A. It lands safely on the airfield.

      B. It crashes and goes down.

C. It loses its instrument panels.

D. It gets a slight vibration in the pedals.

 4-K-4

Task 6   Watch and answer the questions.

1. What happened to the helicopter?

2. What orders were given to aviators and ground forces respectively?

Task 7   Watch again and complete the conversation.

A We got a Black Hawk down. We got a Black Hawk down. Super 61 is down. We (1) _____. Super 61 is (2) _____.

B Get an MH-six on site. (3) _____. (4) _____ the SAR bird. I want (5) _____ and (6) _____ around that (7) _____. Can you (8) _____ in there?

A Roger that.

B Well, (9) _____. The whole city will be coming down on top of them.

A Star 41, (10) _____ and (11) _____ at the crash site.

C Roger that. Star 41 inbound.

A Juliet 25, (12) _____, (13) _____. Women and children among them. Over.

Task 8   Talk and present.

Suppose a company of special operations forces, conducting a counter-terrorism mission in a remote mountainous region, were attacked by a group of terrorists. In this operation, some soldiers are wounded and you are responsible for command and control support in the MEDEVAC mission. How will you effectively communicate with both aviators and ground forces? Create a short conversation using the expressions you've learned in this lesson.

Lima Further Reading

◇ Text A

A Downed Black Hawk in Somalia

Relief Efforts in Somalia

1. Following the downfall of President Siad Barre in 1991, a civil war broke out in Somalia between the faction supporting Interim President Ali Mahdi Mohamed and that supporting General Mohamed Farah Aidid. The United Nations, in cooperation with the Organization of African Unity (OAU) and other organizations, sought to resolve the conflict. The Secretary-General in 1991 dispatched an envoy to whom all faction leaders expressed support for a United Nations peace role. The United Nations also became engaged in providing humanitarian aid, in cooperation with relief organizations. The war had resulted in nearly 1 million refugees and almost 5 million people threatened by hunger and disease.

2. The Security Council in January 1992 imposed an arms embargo against Somalia. The Secretary-General organized talks between the parties, who agreed on a ceasefire, to be monitored by United Nations observers, and on the protection of humanitarian convoys by United Nations security personnel. In April, the Council established the United Nations Operation in Somalia (UNOSOM I).

3. The relief effort was hampered by continued fighting and insecurity. The Security Council in August decided to deploy some 3,000 additional troops to protect humanitarian aid. But the situation continued to worsen, with aid workers under attack as famine threatened 1.5 million people.

4. The United States in November 1992 offered to organize and lead an operation to ensure the delivery of humanitarian assistance. The Security Council accepted the offer and authorized the use of "all necessary means" to establish a secure environment for the relief effort. The Unified Task Force (UNITAF), made up of contingents from 24 countries led by the United States, quickly secured all major relief centres, and by year's end humanitarian aid was again flowing. UNOSOM remained responsible for protecting the delivery of assistance

and for political efforts to end the war.

Things Did Not Go Well

5. There was never even supposed to be a Battle of Mogadishu. In one of his final acts after losing the 1992 election to Bill Clinton, President George H. W. Bush sent American forces into Somalia on a humanitarian mission to bring food to the victims of a raging civil war and man-made famine.

6. But by the fall of 1993, the mission had expanded to one of restoring a government in Somalia. On Oct. 3, a special ops team was sent into Mogadishu to arrest two top lieutenants of the warlord Mohammed Aidid, who controlled the city.

7. "They estimated it would take 30 minutes to 45 minutes to conduct the raid, but things did not go well," says journalist Mark Bowden, who reported on the events of that day.

8. His account, first in *The Philadelphia Inquirer*, then in a book and finally in a blockbuster film, gave the Battle of Mogadishu the name by which it's better known today: Black Hawk Down.

9. Bowden interviewed the men who survived the mission, including Shawn Nelson, an M60 gunner who roped down to the scene from a helicopter.

10. "We immediately started taking fire from the ground. I could see people below us with weapons maneuvering about," he told Bowden.

11. Nelson said that rangers did arrest their two targets, along with about 20 other Somalis who were in a house with them. But taking on so much fire in the busy streets, there was no way to get out fast.

12. "The longer they stayed, the intensity of the fire that the troops encountered increased, including the fire directed at the helicopters overhead," Bowden says.

13. About 40 minutes into the mission, one of the Black Hawk helicopters circling overheard was hit by a rocket-propelled grenade, spun out of control and crashed. Not long after, a second Black Hawk was shot down. More men were sent in to secure the crash sites and get the soldiers out. But the rescue team itself got pinned down.

14. "I said a little prayer," says Spc. Phil Lepre, who was on that rescue convoy, "took off my helmet, looked at my daughter's picture, I said, 'Babe, I hope you have a wonderful life.'"

15. The 15-hour battle that ensued left 18 Americans dead and 73 injured.

Hundreds, perhaps thousands, of Somalis were killed. U. S. Army pilot Mike Durant was captured and held by Somali militants for 11 days.

16. Meanwhile, back in America, the same news networks that broadcast the start of the peaceful humanitarian mission less than a year earlier now ran horrific footage of Aidid supporters desecrating the corpses of U. S. soldiers.

17. All of this intensified the pressure on then-President Clinton to get U. S. troops out of the country.

18. "We had gotten to a point... where we kind of thought that we could intervene militarily without getting hurt, without our soldiers getting killed. The incident that I call Black Hawk Down certainly disabused us of that," Bowden tells Arun Rath, host of "All Things Considered."

If It Happened Again

19. Since 1993, there have been significant advances to America's special operations.

20. "Our ability to gather intelligence to find people, to observe them from a distance with the addition of a fleet of drones that we now have flying is vastly improved," Bowden says. "And we also have special operators who — after Iraq and Afghanistan — who have had more experience conducting the kind of raid that took place back in 1993 than any force like it in the history of the world."

21. If conducted today, the Mogadishu raid would have been done more efficiently, Bowden suspects. He says there also would be better intelligence about the risks ahead of time. But that's not to say there wouldn't be hiccups.

22. "The men who conducted that raid were extremely professional, and they didn't do anything wrong," he says. "The fact is that when you go into combat, it's very not only possible but very likely that... unanticipated things will happen and you'll end up in a much bigger fight than you would prefer."

Task 1   Read the text and answer the following questions.

1. What was the original mission of American forces sent into Somalia in 1992?

2. What were some challenges faced by the soldiers during the raid in Mogadishu according to survivor Shawn Nelson?

Task 2   Read the text and answer true (T) or false (F).

1. The United Nations Operation in Somalia (UNOSOM I) was established to

monitor a ceasefire and protect humanitarian convoys. _____

2. The relief effort in Somalia was successful without any additional military support. _____

3. The United States led a multinational force to secure the delivery of humanitarian aid in Somalia. _____

4. The Unified Task Force (UNITAF) was made up of troops from only the United States. _____

Task 3 Read the text and choose the best answer.

(   )1. What was the main reason the U.S. sent forces to Somalia in 1992?

A. To bring food to people in need.

B. To fight the Somali warlord, Mohammed Aidid.

C. To restore a government in Somalia.

D. To rescue U.S. Army pilot Mike Durant.

(   )2. What happened when one of the Black Hawk helicopters was hit?

A. It landed safely on the ground.

B. It continued flying without any problems.

C. It spun out of control and crashed.

D. It was shot down by the Somali militants.

(   )3. How long were the U.S. troops supposed to be in Mogadishu to arrest the two targets?

A. 15 hours

B. 11 days

C. 30 to 45 minutes

D. About 40 minutes

(   )4. What happened to U.S. Army pilot Mike Durant during the battle?

A. He was able to escape unharmed.

B. He was killed in the battle.

C. He was injured but not captured.

D. He was captured and held by the Somali militants.

(   )5. How have America's special operations improved since 1993?

A. They have better equipment and technology.

B. They have more experienced soldiers.

C. They have more drones and intelligence-gathering capabilities.

D. All of the above.

(   )6. If a similar mission were conducted today, what would be the key differences?

    A. The mission would be carried out with a larger force and greater firepower.

    B. The mission would be planned and executed more efficiently.

    C. The mission would be canceled due to the high risks involved.

    D. The mission would be carried out in the same way as in 1993.

Task 4   Work in groups to discuss the reasons for the failure of this operation.

◇ **Text B**

Deeply Impressed by Chinese Peacekeepers in Africa

Over the past more than 10 years, I have participated in 3 UN peacekeeping missions, namely, the UN Organization Stabilization Mission in the Democratic Republic of the Congo (MONUSCO), the African Union-United Nations Hybrid Operation in Darfur (UNAMID) and the UN Mission in the Republic of South Sudan (UNMISS). In the process, I experienced many difficulties and dangers in peacekeeping operations and witnessed many unforgettable moments of Chinese peacekeepers.

In 2014, due to the outdated equipment of a certain country's peacekeeping helicopter unit, safety accidents occurred from time to time. For this reason, the unit was dismissed by the UN. The UNAMID urgently needed a new peacekeeping helicopter unit. Although several countries had expressed their willingness to deploy such a troop in Darfur, the Sudanese government refused them all for security concern. The lack of helicopters resulted in a near-paralysis of air operations and significant constrains on military operations, which the UNAMID was unable to afford.

In September 2015, when addressing a summit on UN peacekeeping, Chinese President Xi Jinping declared that China would deploy its first helicopter unit for UN peacekeeping missions in Africa. The news became the headline of the day in the UNAMID.

Soon, this China's first peacekeeping helicopter unit was really deployed in Darfur. The Sudanese government embraced for this decision and said it believes in Chinese government's motivation for peace in Darfur. It is understandable that the Sudanese government takes a particularly meticulous stance on the deployment of peacekeeping helicopter unit. In this circumstance, the trust it has shown for China seems to be more precious.

In the morning briefing when the news was declared in UNAMID, the commander on duty called the Chinese military "a force that truly defends peace" and expressed hope for early deployment of China's helicopter unit. His words sparked thunderous applause. As the only Chinese witnessing the moment, I was greeted with admiration and compliment in other attendees' eyes, which I know were not for me, but for the Chinese army and my great motherland.

In 2018, I was appointed the sector commander of the UNMISS. Soon after I took office, Chinese peacekeepers were put under the spotlight abruptly for their efforts in South Sudan, which also made a Chinese saying go popular.

After years of civil wars, many homeless refugees in South Sudan gathered in more than 20 refugee camps, big and small, where they subsisted on living reliefs and safety guarantee provided by the UN. Although the local security environment has improved, few refugees were willing to leave those camps, resulting in an increasing number of inhabitants.

In response, Chinese peacekeepers used their domestic experience and creatively launched a "military-civilian cooperation program" that offered free work skills training for youth in refugee camps, such as home appliance repair and simple mechanical maintenance. With such skills, those young people could find stable jobs in cities and would move out of refugee camps and even could take their families away.

Previously, the UNMISS took a lot of measures to encourage refugees to live on their own, but earned very few results. However, the program launched by Chinese peacekeepers has been proven very successful. It has also inspired peacekeeping forces from other countries to work more effectively in helping refugees restore normal life.

At a high-level meeting of the UNMISS, I explained the reasons behind the effective approach taken by Chinese peacekeepers. "As a Chinese saying goes, 'Give a man a fish, and you feed him for a day. Teach a man to fish, and you feed him for a lifetime.' As young people grasp the skills to make a living, they will put down their weapons and devote themselves to production."

This idea was highly acclaimed by attendees. After the meeting, many colleagues told me that this seemingly simple proverb contains rich Chinese wisdom, which should be the goal upheld by UN peacekeeping missions.

The proverb "Give a man a fish, and you feed him for a day. Teach a man to fish, and you feed him for a lifetime" quickly spread from mouth to mouth, becoming a buzz phrase. Then, the approach of Chinese peacekeepers was further spread, helping the UNMISS win understanding and support from the government and public of South Sudan. Ancient Chinese wisdom and the development concept of contemporary China are implemented in the hinterland of Africa, and have won widespread recognition for their fruitful results. Every time I think of it, I feel very proud as a Chinese.

The experience of participating in peacekeeping missions in Africa for three times over the past more than 10 years has enabled me to witness more and more Chinese military members go abroad with greater confidence to make contributions to world peace and development. At the same time, I am deeply impressed by the sincere expectation, admiration and respect of the UN, the host countries and their people for Chinese peacekeepers. I believe that as a force to safeguard peace, Chinese peacekeepers will continue making greater contributions and accomplishments in the cause of UN peacekeeping with their wisdom and strength.

(The author Ser. Col. He Xing is the director of the teaching and research office for international peacekeeping at the College of International Studies, PLA National University of Defense Technology.)

Task 5   Read the text and choose the best answer.

(   )1. According to the passage, what was the main reason for the dismissal of the peacekeeping helicopter unit from a certain country?

A. Safety concerns

B. Outdated equipment

C. Lack of funding

D. Lack of qualified pilots

(   )2. Why was the Sudanese government initially hesitant to allow the deployment of a new peacekeeping helicopter unit in Darfur?

A. Security concerns

B. Preference for ground troops

C. Distrust of the UN

D. Lack of infrastructure

(　　)3. How did the Sudanese government react to China's decision to deploy its first peacekeeping helicopter unit in Darfur?

A. The Sudanese government remained skeptical and refused the deployment.

B. The Sudanese government expressed concerns about the unit's capabilities.

C. The Sudanese government welcomed the decision and showed trust in China.

D. The Sudanese government negotiated terms for the unit's deployment.

(　　)4. According to the passage，what was the key insight behind the Chinese peacekeepers' approach in South Sudan?

A. Empowering refugees with skills to become self-reliant and independent.

B. Providing long-term financial support to the refugee camps.

C. Encouraging refugees to leave the camps and return to their homes.

D. Collaborating with the local government to address the root causes of the conflict.

Glossary

Chapter One Who We Are: History and Culture

Alpha-Overview

arduous	[ˈɑːdjʊəs]	adj. 艰苦的，艰难的
sublime	[səˈblaɪm]	adj. 崇高的，宏伟的，壮丽的
entrust	[ɪnˈtrʌst]	v. 委托，托付，交代（任务等）
highlight	[ˈhaɪˌlaɪt]	v. 突出，以……为主要装备
electronic warfare		电子战
Union	[ˈjuːnɪən]	美国南北战争中"美利坚合众国"
Confederate	[kənˈfedərət]	美国南北战争中"美利坚联盟国"
hydrogen	[ˈhaɪdrədʒən]	n. 氢，氢气
light aircraft (= light airplane/ light plane)		轻型飞机，小型飞机，轻型航空器
fire correction		射击修正
airlift	[ˈeəlɪft]	n. 空运
evacuation	[ɪˌvækjʊˈeɪʃ(ə)n]	n. 后送，撤退，撤运
aircrew	[ˈeəkruː]	n. 空勤人员，（空军）机组人员
PAAC (PLA Army Aviation **Corps**)	[kɔː(r)]	中国人民解放军陆军航空兵
attack helicopter		攻击直升机；武装直升机
service helicopter		勤务直升机
diversified	[daɪˈvɜːsɪfaɪd]	adj. 多样化的
synergy	[ˈsɪnədʒɪ]	n. 最优方案，协同作用

Bravo-Brief History of Aviation

heritage	[ˈherɪtɪdʒ]	n. 遗产，传统
Signal Corps **Aeronautical** Division	[ˌeərəˈnɔːtɪkl]	通信兵航空处
acquisition	[ˌækwɪˈzɪʃ(ə)n]	n. 采购，获得
merge	[mɜːdʒ]	v. 合并
organic	[ɔːˈɡænɪk]	adj. 建制的
spotter helicopter	[ˈspɒtə(r)]	n. 校射直升机
Operation Just Cause		正义事业行动

Operation Desert Storm		沙漠风暴行动
air observation		对空观察,空中观察
field artillery		野战炮兵
asset	['æset]	n. 资产,部队装备
nucleus	['njuːkliəs]	n. 核心
centerpiece	['sentəpiːs]	n. (最佳的、最具吸引力的)部分
airframe	['eəˌfreɪm]	n. 飞机机体;机身
streamlined	['striːmlaɪnd]	a. 流线型的,效率高的
hard-hitting	[ˌhɑːd 'hɪtɪŋ]	a. 强有力的
mobility	[məʊ'bɪləti]	n. 机动性
fluid	['fluːɪd]	a. (形势)易变的
surveillance	[sɜː'veɪləns]	n. 监视,监督,侦察

Charlie-Core Competencies

MTC (movement to contact)		接敌运动
divert	[daɪ'vɜːt]	v. 牵制,分散(注意力、火力),转向
seize the **initiative**	[ɪ'nɪʃətɪv]	夺取主动权
envelop	[ɪn'veləp]	v. 包围,包封
munition	[mjuː'nɪʃn]	n. 弹药,军火,军需品
ammunition	[ˌæmjʊ'nɪʃn]	n. 弹药,军火
en-route	[ˌɒn'ruːt]	adj. 在途中的,航线上的
PR (personnel recovery)		人员营救
distribution	[ˌdɪstrɪ'bjuːʃn]	n. 分发,分配;分布
survivability	[səˌvaɪvə'bɪlɪti]	n. 生存能力,存活率
COA (course of action)		行动过程,作战进程;作战方案
checkpoint	['tʃekpɒɪnt]	n. 检查站
snow-capped	['snəʊ kæpt]	adj. 白雪覆盖的,白雪皑皑的
plateau	['plætəʊ]	n. 高原
dispatch	[dɪ'spætʃ]	n. 派遣,调度
screen	[skriːn]	v. 掩蔽,遮蔽,屏护
round-the-clock		昼夜不断的,昼夜持续的
projection	[prə'dʒekʃn]	发射,投射

air landing		机降（区别于用降落伞的空降，通过空中运输并在直升机悬停时降落）
armed helicopter		武装直升机
air fleet		航空队，机队
well-armed		全副武装的，装备精良的
fall in		集合
course	[kɔ:s]	*n*. 航向；航线
real-combat		实战的
formation flight		编队飞行
onboard	[ˈɒnˈbɔ:d]	*adv*. 在飞机上
cabin	[ˈkæbɪn]	*n*. 座舱
altitude	[ˈæltɪtjuːd]	*n*. 海拔高度
rappel	[ræˈpel]	*v*. 绕绳下降，索降

Delta-Operational Environment

project	[ˈprɒdʒekt]	*v*. 投送，投射（兵力）
theater	[ˈθɪətə(r)]	*n*. 战区，战场
updraft	[ˈʌpdrɑːft]	*n*. 上升气流
downdraft	[ˈdaʊndrɑːft]	*n*. 下沉气流
turbulence	[ˈtɜːbjələns]	*n*. 紊流，湍流
illumination	[ɪˌluːmɪˈneɪʃn]	*n*. 照明
lethality	[liːˈθælɪti]	*n*. 杀伤力，致命性
versatility	[ˌvɜːsəˈtɪləti]	*n*. 多功能性，多面性，通用性
visibility	[ˌvɪzəˈbɪləti]	*n*. 能见度
vegetation	[ˌvedʒəˈteɪʃn]	*n*. 植被
canopy	[ˈkænəpi]	*n*. 驾驶舱，树冠
concealment	[kənˈsiːlmənt]	*n*. 隐蔽，掩蔽，伪装
vulnerability	[ˌvʌlnərəˈbɪləti]	*n*. 脆弱性
mounted	[ˈmaʊntɪd]	*adj*. 乘车的，骑马的
dismounted	[dɪsˈmaʊntɪd]	*adj*. 徒步的
makeup	[meɪkʌp]	*n*. 组成
pandemic	[pænˈdemɪk]	*adj*. （疾病）大规模流行的

disinfection	[ˌdɪsɪnˈfɛkʃən]	n. 消毒，灭菌
host-nation		东道国
CBRNE （chemical，biological，**radiological**，nuclear，and high-yield explosives）	[ˌreɪdɪəˈlɒdʒɪkəl]	化学、生物、放射性、核和高当量爆炸物的，化生放核爆的
post-landing		降落后的，着陆后的
combat formation		战斗编队，战斗队形
reconnoiter	[ˌrekəˈnɒɪtə(r)]	v. 侦察
infamous	[ˈɪnfəməs]	adj. 声名狼藉的
ill-fated	[ɪl ˈfeɪtɪd]	adj. 结局悲惨的，命运多舛的
handicap	[ˈhændɪkæp]	n. 障碍，不利因素
rotor	[ˈrəʊtə(r)]	n. 旋翼，旋转体
brown-out		沙盲
fuselage	[ˈfjuːzəlɑːʒ]	n.（飞机）机身
slice	[slaɪs]	v. 切，割
peer threat		劲敌（威胁）
guided warhead		制导弹头
unguided warhead		非制导弹头
anti-aircraft artillery		高射炮，高射炮兵
man-portable		便携的
SAM（surface-to-air missile）		地对空导弹，舰对空导弹
penetrate	[ˈpenətreɪt]	v. 突防，突破，贯穿，侵彻

Echo-Further Reading

sophisticated weapon		尖端武器
reviewing stand		检阅台
trail-blazer		开路先锋
storm	[stɔːm]	v. 猛攻，强击
charge	[tʃɑːdʒ]	n. 冲锋，攻击
spurt	[spɜːt]	v. 喷吐，喷射
standstill	[ˈstændstɪl]	n. 停顿，停止
water lock		水闸

mighty	['maɪti]	adj. 强大的；巨大的
bullet	['bʊlɪt]	n. 子弹
fortification	[ˌfɔːtɪfɪ'keɪʃn]	n. 防御工事、筑垒
turret	['tʌrət]	n. 炮塔，旋转枪架
stealth	[stelθ]	n. 隐蔽，隐形，秘密行动
hail-thick		密如雨点的
projectile	[prə'dʒektaɪl]	n. 炮弹，弹丸
dispersion	[dɪ'spɜːʃn]	n. 散开，分散
flight firing		空中发射
trench	[trentʃ]	n. 堑壕，战壕，地沟
dugout	['dʌɡaʊt]	n. 防空洞，防空壕
blindage	['blaɪndɪdʒ]	n. 掩体，隐蔽部
lossratio	['reɪʃɪəʊ]	损失比
squadron	['skwɒdrən]	n. （空军或海军的）中队
sortie	['sɔːti]	n. （出动）架次
Tube-launched **Optically**-tracked Wire-guided （TOW） anti-tank missile	['ɒptɪkəli]	"陶"式反坦克导弹
motorizcd infantry		摩托化步兵
on-foot infantry		徒步步兵
hedgehopping	['hedʒhɒpɪŋ]	adj. 超低空的
leap-frog		蛙跳
hammer-and-**anvil**	['ænvɪl]	铁锤与铁砧
toughened feet	['tʌfnd]	铁脚板
air cavalry		空中骑兵
severity	[sɪ'verəti]	n. 严峻性
disarmament	[dɪs'ɑːməmənt]	n. 裁军，削减军备
epoch-making	['iːpɒk]	划时代的
steed	[stiːd]	n. 战马
Theater Command		战区
Outline of Military Training		军事训练大纲
OMTE （Outline of Military Training and Evaluation）		军事训练与评估大纲

electromagnetic	[ɪˌlektrəʊmæɡˈnetɪk]	*adj.* 电磁的
deep assault		纵深突击
interference	[ˌɪntəˈfɪərəns]	*n.*（无线电信号的）干扰，干涉
Training Section		作训科
spectrum	[ˈspektrəm]	*n.* 光谱，谱
jamming	[ˈdʒæmɪŋ]	*n.* 干扰
paralyze	[ˈpærəlaɪz]	*v.* 使瘫痪
steppe	[step]	*n.* 草原
sprawl	[sprɔːl]	*n.*（城市）杂乱无序拓展的地区

Chapter Two   How We Are Organized: Military Organization

Alpha-Overview

maneuver advantage	[məˈnuːvə]	机动优势
warfighting effectiveness		作战效能
exponential increase	[ˌekspəˈnenʃl]	指数增长
harness	[ˈhɑːnɪs]	v. 控制并利用
battlefield **leverage**	[ˈliːvərɪdʒ]	战场影响力
MTOE（modified table of organization and equipment）		编制装备修正表
CAB（combat aviation brigade）		陆航作战旅
ECAB（expeditionary combat aviation brigadc）		陆航远征作战旅
TAB-A（theater aviation brigade（assault））		战区陆航旅（突击）
TAB-GS（theater aviation brigade（general support））		战区陆航旅（全般支援）
TAOG（theater airfield operations group）		战区机场作业大队
TASMG（theater aviation sustainment maintenance group）		战区航空保障维修大队
ACS（air cavalry squadron）		空中侦察突击中队，空中骑兵中队
AB（attack battalion）		攻击营
AHB（assault helicopter battalion）		突击直升机营
GSAB（general support aviation battalion）		全般支援航空营
ASB（aviation support battalion）		陆航支援营
SSB（security and support battalion）		警戒与支援营
AOB（airfield operations battalion）		机场作业营
FW（fixed-wing）		固定翼
aviation brigade		陆航旅
ASTF（aviation squadron task force）		陆航特遣中队
ABTF（aviation battalion task force）		陆航特遣营
joint aviation		联合航空

Bravo-Aviation Brigades and Enabling Aviation Groups

synchronize	[ˈsɪŋkrənaɪz]	v. 协调，同步
modular	[ˈmɒdjələ(r)]	adj. 模块化的，有标准组件的
tailorable	[ˈteɪlərəbl]	adj. 灵活调整的
DSCA (defense support of civilian authorities)		民事当局支援
HHC (headquarters and headquarters company)		总部及总部连
Gray Eagle company		"灰鹰"（无人机）连
C2 (command and control)		指挥与控制
position personnel		配置人员
augment	[ɔːgˈment]	v. 加强
ATS (air traffic services)		空中交通勤务
AO (area of operations)		作战区域
airfield management operation		机场管理作业
aviation sustainment maintenance		机务，保障维修
limited **depot** sustainment support	[ˈdepəʊ]	有限的军需库保障支援
theater level		战区级
components/end-items		部件/成品
National Maintenance Program		国家维修计划
headquarters **detachment**	[dɪˈtætʃmənt]	司令部分遣队
ASC (aviation support company)		陆航支援连
group support company		大队支援连
fly into **fray**	[freɪ]	飞入战场
the 10th Mountain Division		美国陆军第 10 山地师
MEDEVAC (medical **evacuation**)	[ɪˌvækjʊˈeɪʃ(ə)n]	医疗后送
mass fire	[mæs]	密集射击
ARB (attack **reconnaissance** battalion)	[rɪˈkɒnɪsns]	攻击侦察营
CASEVAC (**casualty** evacuation)	[ˈkæʒuəlti]	伤后送
ground recovery operation		地面救援行动
health support operation		健康支援行动
heavy attack reconnaissance squadron		重型攻击侦察中队
light fighter		轻型战斗机
staff planning		人员规划

contingency operation	[kənˈtɪndʒənsi]	应急行动
ARNG（Army National Guard）		陆军国民警卫队
complement	[ˈkɒmplɪmənt]	n. 补充物
back shop		修理厂
sheet metal blade		金属薄板叶片
Full Kitting		全套（维修）
quality control shop		质量控制车间
defuel	[diːˈfjʊəl]	v. 放油
stockpile	[ˈstɒkpaɪl]	n. 库存
MOC（maintenance operational check）		维修作业检查
multitasking	[ˌmʌltiˈtɑːskɪŋ]	n. 多重任务处理

Charlie-Aviation Battalions and Squadrons

aviation squadron		航空兵中队
USAR（United States Army Reserve）		美国陆军预备役
SWTs（scout weapons teams）		侦察武器小组
AWTs（attack weapons teams）		攻击武器小组
AMT（aviation maintenance troop）		航空维修连
AMC（aviation maintenance company）		航空维修连
FSC（forward support company）		前方支援连
MOS（Military Occupational Specialty）		军事专业
augmentation	[ˌɔːgmɛnˈteɪʃ(ə)n]	n. 增援,加强
contact		n. 接敌
troop	[truːp]	n. 骑兵连
replenishment	[rɪˈplenɪʃmənt]	n. 补充,补给
SME（subject matter expert）		主题专家
subordinate units		下属单位
configuration	[kənˌfɪgəˈreɪʃn]	n. 结构
recast	[ˌriːˈkɑːst]	v. 重组
designate	[ˈdezɪgneɪt]	v. 命名,任命,指定,指派
activate	[ˈæktɪveɪt]	v. 正式成立部队,组建（军事团体）;启动

Delta-Further Reading

night vision goggle		夜视镜
Operation Urgent Fury		暴怒行动
Operation Earnest Will		坚定决心行动
Operation Mount Hope Ⅲ		希望山三号行动
forward looking **infrared** device	[ˌɪnfrəˈred]	前视红外装置
retrieve	[rɪˈtriːv]	v. 找回
spearhead	[ˈspɪəhed]	v. 做……的先锋;带头做;领先突击
insertion	[ɪnˈsɜːʃn]	n. 插入;投送
pioneer	[ˌpaɪəˈnɪə(r)]	v. 开创;做先锋
migrate	[maɪˈɡreɪt]	v. 转移
collocate	[ˈkɒləkət]	v. 搭配;共同驻扎
posture	[ˈpɒstʃə(r)]	v. 姿态;构造
three-time volunteer		三阶段训练志愿兵
in a **nutshell**	[ˈnʌtʃel]	简而言之
line unit		基层单位,一线部队
resource	[rɪˈsɔːs]	v. 向……提供资金(或设备)
permeate	[ˈpɜːmieɪt]	v. 贯穿
maintainer	[meɪnˈteɪnə]	n. 维修人员
Transportation Corps		运输兵团
USAALS(US Army Aviation Logistics School)		美国陆军航空后勤学校
initial-entry		首次进入
Advanced Noncommissioned Officer Course		高级士官课程
Warrant Officer Aviation Maintenance **Technician** Course	[tekˈnɪʃn]	准尉航空维修技师课程
Aircraft **Armament** Maintenance Technician Course	[ˈɑːməmənt]	飞机武器维修技师课程
pneudraulic repairer	[njuːˈdrɔːlɪk]	气动维修工
avionics repairer	[ˌeɪviˈɒnɪks]	航电维修工
Warrant Officer Technician Basic and Advanced Courses		准尉技师基础和高级课程
powerplant	[ˈpaʊəˌplɑːnt]	n. 动力装置;动力系统
powertrain	[ˈpaʊəˌtreɪn]	n. 动力总成,动力系统

WOTD（Warrant Officer Training Division）		准尉训练部
WOBC（Warrant Officer Basic Course）		准尉基础课程
capstone	['kæpstəʊn]	n. 顶点
austere	[ɒ'stɪə(r)]	adj. 严峻的
Combat Aviation Brigade Maintenance Meeting		陆航作战旅维修会议
cadre	['kɑːdə(r)]	n. 干部,骨干
WOAC（Warrant Officer Advanced Course）		准尉高级课程
CTSSB（Critical Task and Site Selection Boards）		关键任务和选址委员会
Aircraft Powertrain Repairer		飞机动力系统维修工
Aircraft Powerplant Repairer		飞机动力装置维修工
Aircraft Pneudraulics Repairer		飞机气动维修工
Avionic Mechanic		航空机械师
Aircraft **Electrician**	[ɪˌlek'trɪʃn]	飞机电工
AIT（Advanced Individual Training）		高级单兵训练
SALUTE（size，activity，location，unit，time and equipment）report		SALUTE(尺寸、活动、位置、部队、时间和设备)报告
Call for Fire/Adjust Fire Report		呼叫炮火/矫正炮火报告
Aviation Warfighter		陆航战士

Chapter Three   What We Fly: Helicopters and Drones

Alpha-Overview

payload	['peɪloʊd]	n. 有效载荷
multirole	[ˌmʌltɪ'rəʊl]	adj. 多作用的,通用的
utility helicopter		通用直升机
transport helicopter		运输直升机
search and rescue helicopter		搜救直升机
missile	['mɪs(ə)l]	n. 导弹
hardpoint	['hɑːdpɒɪnt]	n. 承力点
aerial	['erɪəl]	n. 航空的,空中的
cargo	['kɑːrgəʊ]	(船或飞机装载的)货物
underslung	[ˌʌndər'slʌŋ]	下悬式的
mammoth	['mæməθ]	n. 猛犸,庞然大物; adj. 巨大的,庞大的
air assault		空中突击
reconnaissance	[rɪ'kɒnɪsns]	n. 侦察
battlefield	['bætlfiːld]	n. 战场
versatile	['vɜːrsət(ə)l]	adj. 多用途的,多功能的
minefield	['maɪnfiːld]	n. 布雷区
durable	['dʊrəb(ə)l]	adj. 持久的,耐用的
agile	['ædʒ(ə)l]	adj. 敏捷的,灵活的
avionics	[ˌeɪvi'ɑːnɪks]	n. 航空电子设备
vital	['vaɪt(ə)l]	adj. 生命的,维持生命所必需的
self-defence	[ˌself dɪ'fens]	n. 自卫
infrared	[ˌɪnfrə'red]	adj. 红外线的,(设备,技术)使用红外的
jammer	['dʒæmər]	n. 干扰发射机
laser detection		激光探测

Bravo-Brief History of Helicopters

battlefield interdiction		战场封锁
equipment system		装备体系
training helicopter		教练直升机
military budget		军费预算
test flights		试飞
horsepower	[ˈhɔːspaʊə(r)]	n. 马力
collective pitch		主旋翼螺距
tail rotor		尾旋翼
underpowered	[ˌʌndəˈpaʊəd]	adj. 动力不足的
dynamic stress		动应力
material failure		材料失效
equipment failure		设备失效
raid	[reɪd]	n. 突袭
design features		设计特点
radar cross section		雷达散射截面
spot	[spɒt]	v. 发现
operate	[ˈɒpəreɪt]	v. 运转
detection	[dɪˈtekʃn]	n. 侦察, 发现
radar	[ˈreɪdɑː(r)]	n. 雷达
wreckage	[ˈrekɪdʒ]	n. 残骸
the US Special Forces		美国特种部队
cancel	[ˈkænsl]	v. 撤销, 中止
shield	[ʃiːld]	v. 掩护, 保护
adversary	[ˈædvəsəri]	n. 对手, 敌手

Charlie-Helicopter Structure

aerodynamics	[ˌeərəʊdaɪˈnæmɪks]	n. 空气动力(特性); 空气动力学
tubular	[ˈtjuːbjələ(r)]	adj. 管状的
skid	[skɪd]	n. 滑橇
landing gear		起落架
retractable	[rɪˈtræktəbl]	adj. 可伸缩的, 可缩回的
reciprocating engine		活塞式发动机, 往复式发动机

turbine engine		涡轮发动机
rotor system		旋翼系统，转子系统
mast	[mɑːst]	n. 旋翼轴，桅杆
hub	[hʌb]	n. 旋翼桨毂
rotor blade		螺旋桨，旋翼叶片
hinge	[hɪndʒ]	n. 铰链
Antitorque System		反扭矩系统
torque	[tɔːk]	n. (尤指机器的)扭转力，转(力)矩
countertorque	['kaʊntətɔːk]	n. 反扭矩，反力矩
modulate	['mɒdjəleɪt]	v. 调整，调节
throttle	['θrɒtl]	n. 节流阀，油门，手控加速器(手油门)
airfoil	['eəfɔɪl]	n. 翼面，翼形
thrust	[θrʌst]	n. 推力
drag	[dræg]	n. (空气)阻力
cockpit	['kɒkpɪt]	n. (飞机、小船或赛艇的)驾驶室
spinning blades		旋转的叶片
propeller	[prə'pelə(r)]	n. (飞机或轮船的)螺旋桨，推进器
fantail	['fænteɪl]	n. 扇形尾翼
inflict	[ɪn'flɪkt]	v. 造成，施以，加害
zero in on		专注于，瞄准，对准
maneuver	[mə'nuːvə(r)]	v. 操纵，调遣，机动
hydraulic system		液压系统
fine-tune	[ˌfaɪn'tjuːn]	v. 优化调整；对……微调
hover	['hɒvə(r)]	v. 悬停
longbow Apache		长弓阿帕奇直升机

Delta-Flight Preparations

airworthy	['eəwɜːði]	n. 适航的，耐飞的
align with		与……一致，校准
perpendicular	[ˌpɜːpən'dɪkjələ(r)]	adj. 垂直的，成直角的
parameter	[pə'ræmɪtə(r)]	n. 界限，范围；参数，变量
deplane	[diː'pleɪn]	v. 下飞机，离开飞机

practicable	['præktɪkəbl]	*adj*. 可行的，有用的，适用的
harness strap		安全带
night flare		夜间照明弹
heliograph	['hiːlɪəɡrɑːf]	*n*. 日光反射信号器，日光仪
antiseptic	[ˌæntiˈseptɪk]	*adj*. 抗菌的，无菌的，消过毒的
plaster	['plɑːstə(r)]	*n*. 膏药
strobe	[strəʊb]	*n*. 闪光灯
strop	[strɒp]	*n*. 带索，环索，吊带
sachet	['sæʃeɪ]	*n*.（密封的塑料或纸质）小袋
discrepancy	[dɪˈskrepənsi]	*n*. 差异，矛盾
pertinent	['pɜːtɪnənt]	*adj*. 相关的，有关的；相宜的
cabin door		客舱门，机舱门
fire extinguisher		灭火筒，灭火器

Echo-Attack Helicopters

autocannon	['ɔːtəʊ'kænən]	*n*. 机炮
scout	[skaʊt]	*n*. 侦察
AH-64 Apache		ΛH-64 阿帕奇武装直升机
Ka-52 Alligator		（俄罗斯）卡-52 短吻鳄武装直升机
tailwheel-type		尾轮式
tandem	['tændəm]	*adj*.（飞机）串座式的
TADS（Target Acquisition and Designation System）		目标捕获指示系统
PNVS（Pilot Night Vision System）		飞行员夜视系统
passive infrared countermeasures		无源红外对抗
IHADSS（Integrated Helmet and Display Sight System）		综合头盔显示系统
Panama	['pænəmɑː]	巴拿马
low intensity operation		低强度作战
ferry range		转场航程
service ceiling		飞行高度，［航］实用升限，升高限度
endurance	[ɪnˈdjʊərəns]	*n*. 续航时间，耐力
Longbow radar		长弓雷达

Hydra rocket pod		火蛇火箭吊舱
pierce	[pɪəs]	v. 穿过，透入
chain gun		弹链式枪
round of ammunition		（炮弹、子弹）发
rate of fire		射速
round	[raʊnd]	n. （炮弹、子弹）发
infrared sensor	[ˌɪnfrə'red]	红外传感器
barrel	['bærəl]	n. 枪管
cross hairs	['krɒs heəz]	十字线，瞄准十字线
elbow grease		费力的工作
dome	[dəʊm]	n. 半球形物，圆顶状物

Foxtrot-Utility Helicopters

Sikorsky Aircraft		美国西科斯基飞机公司
Bell CH-1 Iroquois	['ɪrəkwɒɪz]	贝尔 UH-1 直升机
101st Combat Aviation Brigade		第 101 陆航作战旅
field		v. 列装
special operation		特种作战
crew chief		炮长，机长，车长
sling	[slɪŋ]	n. 吊具
minigun		n. 机枪
gatling gun		加特林机枪
laser guided missile		激光制导导弹
Stinger air-to-air missile		"毒刺"空对空导弹
extract	[ɪk'strækt]	v. 撤离
fire suppression		火力压制
GE engine		通用电气发动机
composite	['kɒmpəzɪt]	adj. 合成的，复合的
wide **chord** rotor blade	[kɔːd]	宽弦桨叶
situational awareness		态势感知
nimble	['nɪmb(ə)l]	adj. （行动）灵活的
Saddam Hussein		萨达姆·侯赛因（伊拉克前总统）

SH-60 Seahawk		SH-60"海鹰"
VH-60 White Hawk		VH-60"白鹰"
Marine One		海军陆战队一号
Operation Neptune Spear		海神之矛行动
Al-Qaida		基地组织
Osama bin Laden		奥萨马·本·拉登

Golf-Cargo Helicopters

ubiquitous	[juːˈbɪkwɪtəs]	adj. 普遍存在的；无处不在的
ramp	[ræmp]	n. 斜坡，坡道
haul	[hɔːl]	vi. 拖，拉
rotor aircraft		旋翼机
maintenance	[ˈmeɪntənəns]	n. 保养，维修
vertical-lift aircraft		垂直起落飞机
disaster-relief operations		灾难援助行动
vibration	[vaɪˈbreɪʃn]	n. 震动
aft section		后舱
pylon	[ˈpaɪlən]	n. (飞机)炸弹挂架
terrain	[təˈreɪn]	n. 地形，地域，地带
Chinook	[tʃɪˈnuːk]	n. 支奴干直升机
load	[ləʊd]	n. 装载量
payload capacity		有效载重能力，有效载重量，业载能力
operational capability		作战能力
tactical	[ˈtæktɪkl]	adj. 战术的
armor	[ˈɑːmə(r)]	n. 装甲，盔甲
howitzer	[ˈhaʊɪtsə(r)]	n. 榴弹炮
ammo	[ˈæməʊ]	n. [口]弹药，军火
staggering	[ˈstæɡərɪŋ]	adj. 难以置信的，令人震惊的
lifting capacity		载重量，空运能力
SUSV (Small Unit Support Vehicle)		小型单位支援车
hook up		用钩钩住，将……联接起来
sway	[sweɪ]	v. 使摇动

sling load		吊载,起吊载荷
spin	[spɪn]	v. 旋转
stabilize	['steɪbəlaɪz]	v. (使)稳定,(使)稳固
contingency		n. 偶然发生的事故,意外事故,应急措施

Hotel-Chinese Military Helicopters

modified	['mɒdɪfaɪd]	adj. 改良的,改进的
fenestron	[fə'nestrən]	n. (航空)一种旋翼机的后旋翼
acoustic	[ə'kuːstɪk]	adj. 声音的,听觉的
exhaust	[ɪg'zɔːst]	n. 排气装置,排气管[孔]
MMW (millimeter wave)		毫米波
HMS (helmet mounted sight)		头盔瞄准具
aeronautics	[ˌeərə'nɔːtɪks]	adj. 航空(学)的
astronautics	[æstrə'nɔːtɪks]	n. 太空航空学
optical	['ɒptɪkl]	adj. 光学的
mount	[maʊnt]	v. 配有……
FLIR (Forward looking infra-red)		前视红外
stub wings		短翼
armament	['ɑːməmənt]	n. 武器;军备
turboshaft	['tɜːbəʊʃɑːft]	n. 涡轮轴
trim	[trɪm]	n. (飞机、潜艇的)配平,配平状态
diameter	[daɪ'æmɪtə(r)]	n. 直径
gear	[gɪə(r)]	n. 齿轮,传动装置
utility	[juː'tɪləti]	n. 功用,效用
resemble	[rɪ'zembl]	v. 像……,类似于
subsidiary	[səb'sɪdiəri]	n. 附属机构
maiden flight		首航
military parade		阅兵
formation	[fɔː'meɪʃn]	队形,编队
indigenously	[in'didʒinəsli]	adv. 本土
variant	['veəriənt]	n. 变体,变型

aerodynamic	[ˌeərəʊdaɪˈnæmɪk]	*adj*. 空气动力学的
angular	[ˈæŋɡjələ(r)]	*adj*. 角度的
fairing	[ˈfeərɪŋ]	*n*. 减阻装置，整流罩
incorporate	[ɪnˈkɔːpəreɪt]	*v*. 包含，加上
antisubmarine	[ˌæntɪˌsʌbməˈriːn]	*adj*. 反潜艇的
align	[əˈlaɪn]	*v*.（使）成一条直线
fire support		火力支援
Fly by Wire technology		电传飞控技术
MAWS（Missile Approach Warning System）		导弹逼近预警系统
electronic warfare suite		电子战套件

India-Drones

gimbal	[ˈdʒɪmb(ə)l]	*n*. 平衡环，常平架
get the hang of		得知……的窍门，熟悉某物的用法
sensors	[ˈsensəz]	*n*.［自］传感器，感应器
ISR（Intelligence，Surveillance，and Reconnaissance）		情报，监视，侦察
covert	[ˈkʌvət]	*adj*. 隐蔽的，秘密的
thermal camera		热成像摄影机，热感摄像机
Damascus	[dəˈmæskəs]	*n*. 大马士革（叙利亚首都）
escalation	[ˌeskəˈleɪʃ(ə)n]	*n*. 升级，恶化
Nevada	[nəˈvɑːdə]	*n*. 内华达（美国州名）
tarmac	[ˈtɑːmæk]	*n*. 铺有柏油碎石的飞机跑道

Juliet Chinese Military Drones

medium altitude		中等高度
MQ-9 REAPER		收割者，MQ-9 死神无人机
takeoff weight		［航］起飞重量
SAL（Semi-Active Laser）		半主动激光

Kilo-Further Reading

naming system		命名系统
Marine Corps		海军陆战队
aerospace vehicle		航空航天飞行器，航天飞机
MDS（Mission-Design-Series） designations		"任务—设计—系列"命名方式
model designation		型号
nickname		*n.* 绰号，外号
tribe		*n.* 部落，宗族
intermittently	[ˌɪntərˈmɪtəntli]	*adv.* 断断续续地
surprise attack		*v.* 突袭
sweep		*v.* 扫荡；扫射
heroically	[həˈrəʊɪkli]	*adv.* 英勇地
primitive		*adj.* 原始的
musket	[ˈmʌskət]	*n.* 火枪，滑膛枪，毛瑟枪，步枪
warrior		*n.* 战士
the Medal of Honor		荣誉勋章
official regulation		官方规定
split		*v.* 分离
warfighter		*n.* 作战人员，军人
hoverfly		*n.* 食蚜蝇
dragonfly		*n.* 蜻蜓
instruction		*n.* 指令
flank		*n.* 侧翼
the Great Plain		美国大平原
Army Regulation		陆军条例
criteria		*n.* 标准
dignity		*n.* 尊严
aggressive	[əˈɡresɪv]	*adj.* 好斗的，有进取心的；攻击性的，侵略的
flexibility		*n.* 灵活性
firepower		*n.* 火力
manufacture	[ˌmænjʊˈfæktʃə(r)]	*v.* 批量生产　*n.* 工业品
the Bureau of Indian Affairs		美国印第安事务局

fearsome		*adj.* 可怕的
reptile	[ˈreptaɪl]	*n.* 爬行动物
insect		*n.* 昆虫
cobra	[ˈkəubrə]	*n.* 眼镜蛇
scorpion	[ˈskɒrpiən]	*n.* 蝎子
rescind	[rɪˈsɪnd]	*v.* 废止
ritually	[ˈrɪtʃuəli]	*adv.* 仪式上
combat drone		*n.* 无人战斗机,战斗蜂
pressingly		*adv.* 恳切地;固执地;紧迫地
estrangement	[ɪˈstreɪndʒmənt]	*v.* 远离,疏远
shy away from		*v.* 避免,躲避
downscale		*v.* 缩小……的尺寸(或规模),降低……的程度
diffusion	[dɪˈfjuːʒn]	*v.* 扩散,传播
prestige	[preˈstiːʒ]	*n.* 声望,威信
exemplify	[ɪɡˈzemplɪfaɪ]	*v.* 举例说明
refrain	[rɪˈfreɪn]	*v.* 克制,避免
PTSD (posttraumatic stress disorder)		*n.* 创伤后应激障碍
minimise	[ˈmɪnɪmaɪz]	*v.* 使最小化

Chapter Four   What We Do: Mission and Command

Alpha-Overview

AGO（Air-ground operations）		空地作战
aerial maneuver force		空中机动部队
ground forces		地面部队
aviation operations		航空作战
asymmetric advantage	[ˌeɪsɪˈmetrɪk]	非对称优势
substitute		n. 替代
battlespace		n. 战场，作战空间
counterattack		n. 反攻，反击
shape the battlespace		塑造战场空间
close fight		近战
positional advantage		位置优势
combat operation		作战行动
CS（Combat Support）		作战支援
CSS（Combat Service Support）		作战勤务支援
direct fire		直接火力
standoff precision weapons		防区外精确武器
Security		n. 警戒
Air Movement		空中机动
Aeromedical Evacuation		空中医疗后送
Command and Control Support		指挥与控制支持

Bravo-Radio Communications

brevity	[ˈbrevəti]	n. 简洁，简炼
frequency	[ˈfriːkwənsi]	n.（声波或电磁波振动的）频率
key the transmitter		按发射器上的键
transmit	[trænzˈmɪt]	v. 传输，发射
aircraft **identification**	[aɪˌdentɪfɪˈkeɪʃn]	飞机识别
initial contact		初次联系

call signs		呼号，代号
transmission	[trænz'mɪʃn]	n. 传输，发射
abbreviate	[ə'bri:vieɪt]	v. 缩写，缩略
proword		n. 电台业务通信代字
closing down		暂时关机
mobile from		从……地方出发
out		通话结束，关机不再联系
over		完毕
radio check		信号如何
roger		收到
acknowledge	[ək'nɒlɪdʒ]	（你方）收到回复并照办
affirmative	[ə'fɜːmətɪv]	是/肯定的
negative		不/没有/不同意
message		我方有一条信息
send		请讲
wilco(will comply)	['wɪlkəʊ]	（我方）收到照办，遵命
I spell		我进行拼读
say again		请重复一下
speak slower		请说慢一点
read back		请复述
break		分割通信段落，表示后接下一段内容
loud and clear		声音洪亮且清晰
broken and unreadable		不清楚，无法识别
shack		简陋的小屋，避难所
air evacuation team		空中后送队
stay at the site		留在现场
food distribution center		食物配送中心
Request permission to engage.		请求交战
take fire		开火

Charlie-Attack

MUM-T（Manned-Unmanned Teaming）		有人—无人编队
decisive operation		决胜行动
shaping operation		塑造行动
find，fix，and destroy		寻找、定位和摧毁
echelon	[ˈeʃəlɒn]	n. 梯队，层级
disrupt	[dɪsˈrʌpt]	v. 扰乱，使混乱
tempo	[ˈtempəʊ]	n. 节奏
audacity	[ɔːˈdæsəti]	n. 大胆
simultaneity	[ˌsɪmltəˈneɪəti]	n. 同时
maneuverability	[məˌnuːvərəˈbɪləti]	n. 机动性
standoff		a. 防区外（攻击能力）有距离的
zero in on		瞄准
Saudi Arabia		沙特阿拉伯
intrude	[ɪnˈtruːd]	v. 侵入
Baghdad		巴格达
radio silence		无线电静默
Hellfire		地狱火导弹
terrain-following		地形跟踪，地形显示
combat load		战斗载荷
H-Hour		战斗发起时间
Pave Low		"低空铺路者"特种作战直升机
Doppler		多普勒
SOF（Special Operations Forces）		特种作战部队
click		n. 公里
MiG		米格（战斗机）
wage	[weɪdʒ]	v. 发动
coalition	[ˌkəʊəˈlɪʃn]	n. 联盟
dominance	[ˈdɒmɪnəns]	n. 支配地位
degrade	[dɪˈɡreɪd]	v. 削弱
scramble	[ˈskræmbl]	v. 命令（飞机）紧急起飞
sneak	[sniːk]	v. 偷偷地行动
stream	[striːm]	v. 接续行动

compound	[ˈkɒmpaʊnd]	n. 有围栏（或围墙）的场地（内有工厂或其他建筑群）
oblivious	[əˈblɪvɪəs]	adj. 未察觉的
unleash	[ʌnˈliːʃ]	v. 突然释放
radar dish		雷达天线（反射器）
expend	[ɪkˈspend]	v. 消耗
knock out		摧毁
Flechette		火箭弹的一种型号

Delta-Air Assault

RW（rotary wing）aircraft		旋转翼飞机
mass	[mæs]	v. 集结
dislocate	[ˈdɪsləkeɪt]	v. 扰乱，使混乱
TCF（Tactical Combat Force）		战术战斗部队
penetration	[ˌpenəˈtreɪʃn]	n. 穿透
AATF（Army Aviation Task Force）		空中突击特遣队
PZ（Pickup Zone）		搭载区
LZ（Landing Zone）		降落区，着陆区
helo	[ˈheɪləʊ]	直升机（缩写）
Chalk		直升机别称，或直升机上的部队
fast rope		机降、机降用绳
perimeter	[pəˈrɪmɪtə(r)]	n. 边缘，环形防线

Echo-Reconnaissance

relative advantage		相对优势
disposition	[ˌdɪspəˈzɪʃn]	n. 部署
aviation/ground maneuver force		空中/地面机动（作战）部队
zone reconnaissance		地带侦察
area reconnaissance		地区侦察
airhead	[ˈeəhed]	n. 空降场（建于敌占区内用来供应补给和撤退部队及装备的区域）
route reconnaissance		路线侦察

cross country mobility corridor		越野机动走廊
reconnaissance in force		武力侦察
decisive engagement		决战
reinforce	[ˌriːɪnˈfɔːs]	v. 增援
UTM (Universal Transverse Mercator) coordinates		通用横轴墨卡托坐标系
control panel		控制面板
multi sensor		多传感器
fly at nap of the earth altitudes		超低空飞行
MTI (moving target indicator)		活动目标指示器
periscope	[ˈperɪskəʊp]	n. 潜望镜
laser range finder		激光测距仪
forward-looking infrared sensor		前视红外传感器
intelligence center		情报中心

Foxtrot-Movement to Contact

enabling tasks		保障任务
security force		警戒部队
commander's intent		指挥官意图
follow-on forces		后续部队
uncommitted forces		未投入战斗的部队，预备队
premature deployment		过早展开
terrain reconnaissance		地形侦察
doctrinally	[dɑːkˈtraɪnəli]	adv. 教条地
cordon and search	[ˈkɔːdn]	封锁搜查
delay	[dɪˈleɪ]	v. 延迟火力支援
subordinate forces		所属部队
supporting distances		支援距离（部队可及时赶到以支援另一部队的距离）
close air support		近距离空中支援
air interdiction		空中遮断，空中阻滞（切断敌人后方供应线）
counterair operations		防空作战，对空作战

search and attack		搜索与攻击
hit-and-run tactics		扰乱战术
decentralized	[ˌdiːˈsentrəlaɪzd]	v. 分散的
fix	[fɪks]	v. 钳制,钉死,将敌人固定于原地
mounted or dismounted elements		乘车/步行部队
hasty attack		仓促进攻,急速进攻
deliberate attack		周密计划的进攻

Golf-Security

stationary	[ˈsteɪʃənri]	adj. 静止的,不动的
aerial C2		空中指挥与控制
enemy contact		接敌
repel	[rɪˈpel]	v. 击退
touch base		准备降落
clear	[klɪə]	v. 获得(行动)许可
touch down at target		安全着陆
on the deck		已着陆
staging arca		集结区域
on the ground		着陆
hold pattern		空中悬停模式
delivered chalks		准备出动部队
cover pattern		掩护模式
overhead pattern		空中盘旋模式
provide sniper cover		提供空中掩护
RPG（Rocket Propelled Grenade）		便携式火箭弹
a sitting duck		活靶子,易被击中的目标
come out		离开

Hotel-Air Movement

airdrop	[ˈeədrɒp]	n./v. 空投
feasible	[ˈfiːzəbl]	adj. 可行的
configure	[kənˈfɪɡə(r)]	v. 配置

non-combatant	[ˌnɒnˈkɒmbətənt]	n. 非战斗人员
circulation	[ˌsɜːkjəˈleɪʃn]	n. 流动,流通
supplement	[ˈsʌplɪmənt]	v. 补充,增补
coordination	[kəʊˌɔːdɪˈneɪʃn]	n. 协调
register	[ˈredʒɪstə(r)]	v. 显示
Richter	[ˈrɪktə]	里氏
devastate	[ˈdevəsteɪt]	v. 毁灭,摧毁
airland	[ˈeəlænd]	v. 空降
terrace	[ˈterəs]	n. 梯田
relief goods		救援物资
apparent	[əˈpærənt]	adj. 显而易见的
resupply	[ˌriːsəˈplaɪ]	n. 再补给
chase after		追捕
zip up		(拉链)拉上
toss	[tɒs]	v. 扔,抛
scatter	[ˈskætə(r)]	v. 分散
randomly	[ˈrændəmli]	adv. 随机
tempting	[ˈtemptɪŋ]	adj. 诱人的
potshot	[ˈpɒtʃɒt]	n. 盲目射击
armor plating	[ˈɑːmə ˈpleɪtɪŋ]	n. 装甲钢板
go-to	[ˈgəʊ tuː]	adj. 寻求协助的

India-Aeromedical Evacuation

provision	[prəˈvɪʒn]	n. 提供,供给
unregulated	[ʌnˈregjʊleɪtɪd]	adj. 不受管制的
BSA（brigade support area）		旅支援区
evacuate	[ɪˈvækjʊeɪt]	v. 撤离,撤出
advent	[ˈædvent]	n. 出现
Burma	[ˈbɜːmə]	缅甸
dedicated	[ˈdedɪkeɪtɪd]	adj. 专用的
revolutionize	[ˌrevəˈluːʃənaɪz]	v. 彻底改变
medics	[ˈmedɪks]	n. 救护人员,医务人员

mortality	[mɔːˈtæləti]	n. 死亡率
suction	[ˈsʌkʃn]	n. 抽吸
triage	[ˈtriːɑːʒ]	n. 伤员鉴别分类
non-compressible	[ˌnɒnˈkɒmprɪˈsəbl]	adj. 不可压的
abdominal	[æbˈdɒmɪnl]	adj. 腹部的
DARPA （Defense Advanced Research Projects Agency）		美国国防高级研究计划局（美国国防部属下的一个行政机构，负责研发用于军事用途的高新科技）
goggle	[ˈɡɒɡl]	n. 护目镜
medical intervention		医疗干预
carve out		开辟
hang on		坚持

Juliet-Personnel Recovery

unassisted recovery		无协助的营救（自我营救）
immediate recovery		即时营救
deliberate recovery		周密的营救
external supported recovery		外部支持的营救
interagency	[ˌɪntərˈeɪdʒənsi]	adj. 跨机构的
PR officer		营救军官
PR representative		营救代表
at a moment's notice		随时
disembark	[ˌdɪsɪmˈbɑːk]	下（车、船、飞机等）

Kilo-Command and Control Support

communication relay package		通信中继组件
ATS（air traffic services）		空中交通勤务
CAC（command aviation company）		航空指挥连
traverse	[trəˈvɜːs]	v. 跨越
aerial retransmission		空中重传
early entry CP		先期进入指挥所
onboard communications linkages		机载通信链

staff	[stɑːf]	*n.* 参谋人员
rehearsal	[rɪˈhɜːsl]	*n.* 预演
inbound	[ˈɪnbaʊnd]	*adj.* 归航的，到达的

Lima-Further Reading

envoy	[ˈenvɔɪ]	*n.* 使者；代表
hamper	[ˈhæmpə(r)]	*v.* 妨碍
raging	[ˈreɪdʒɪŋ]	*adj.* 猛烈的
famine	[ˈfæmɪn]	*n.* 饥荒
warlord	[ˈwɔːlɔːd]	*n.* 军阀
blockbuster	[ˈblɒkbʌstə(r)]	*n.* 大片
ranger	[ˈreɪndʒə(r)]	*n.* 游骑兵，突击队员
pin sb. down		按住；使动弹不得
ensue	[ɪnˈsjuː]	*v.* 接着发生
MONUSCO		联合国刚果民主共和国稳定特派团
UNAMID		联合国非洲联盟达尔富尔混合特派团
UNMISS		联合国南苏丹特派团
meticulous	[məˈtɪkjələs]	*adj.* 小心谨慎的
hinterland	[ˈhɪntələænd]	*n.* 偏远地区

Acronyms and Abbreviations

AATF air assault task force 空中突击特遣队

ABTF aviation battalion task force 航空(陆航)特遣营

ACM airspace coordinating measure 空域协调措施

ADAM air defense airspace management 防空空域管理

AE aeromedical evacuation 空中医疗后送

AGO air-ground operations 空地作战

AHB assault helicopter battalion 突击直升机营

AHC assault helicopter company 突击直升机连

AMC aviation maintenance company 航空(陆航)维修连

AMT aviation maintenance troop 航空(陆航)维修连

AO area of operations 行动区域、作战区域

AOB airfield operations battalion 机场作业营

AB attack battalion 攻击营

AC attack company 攻击连

ARNG Army National Guard 陆军国民警卫队

ACS air cavalry squadron 空中侦察突击中队,空中骑兵中队

ACT air cavalry troop 空中侦察突击连,空中骑兵连

ASB aviation support battalion 航空(陆航)支援营

ASC aviation support company 航空(陆航)支援连

ASTF aviation squadron task force 航空(陆航)特遣中队

ATC air traffic control 空中交通管制

ATNAVICS air traffic navigation, integration, and coordination system 空中交通导航、综合、协调系统

ATS air traffic services 空中交通勤务

ATSSE air traffic services standardization element 空中交通勤务标准化分队

AWT attack weapons team 攻击武器小组

AXP ambulance exchange point 救护车换乘点

BAE brigade aviation element 旅航空(陆航)分队

BAMO brigade aviation maintenance officer 旅航空(陆航)维修军官

BCT brigade combat team 旅战斗队,作战旅

BDA battle damage assessment 战斗毁伤评估

C2 command and control 指挥与控制

CAB combat aviation brigade 作战航空旅、陆航作战旅

CAC command aviation company 航空(陆航)指挥连

CAS close air support 近距离空中支援

CASEVAC casualty evacuation 伤员后送

CBRNE chemical，biological，radiological，nuclear，and high yield explosives 核生化辐射和高当量炸药

COA course of action 行动方案，行动过程

CP command post 指挥所

DART downed aircraft recovery team 坠机营救小组

DCA defensive counter-air 防御性防空

DS direct support 直接支援

DSCA Defense Support of Civilian Authorities 民事当局国防支援

ECAB expeditionary combat aviation brigade 远征作战航空旅、陆航远征作战旅

EW electronic warfare 电子战

FARP forward arming and refueling point 前置弹药和燃油补给点

FLOT forward line of own troops 己方前锋线

FSC forward support company 前方支援连

FW fixed-wing 固定翼

GPS Global Positioning System 全球定位系统

GS general support 全般支援

GSAB general support aviation battalion 全般支援航空（陆航）营

HA holding area 待机地域

HHC headquarters and headquarters company 总部及总部连

HSS health service support 卫勤保障

IADS integrated air defense systems 一体化防空系统

IED improvised explosive device 简易爆炸装置

IO information operations 信息作战

IPB intelligence preparation of the battlefield 战场情报准备

IR infrared 红外线

ISB intermediate staging base 中间整备基地

JAGIC joint air-ground integration center 联合空地集成中心

JFC joint force commander 联合部队指挥官

LNO liaison officer 联络官

LOA limit of advance 前进界限

LOC line of communications 通信（交通）线

LOS line of sight 视距

LSCO large-scale combat operations 大规模战斗行动

LZ landing zone 着陆区

MANPADS man-portable air defense system 便携式防空系统

MEDEVAC medical evacuation 医疗后送

MTF medical treatment facility 医疗设备

MTOE modified table of organization and equipment 编制装备修正表

MUM-T manned unmanned teaming 有人-无人编队

OCA offensive counter-air 进攻性防空

OE operational environment 作战环境

OP observation post 观察哨所

OPCON operational control 作战控制

PC production control 生产控制

PIR priority intelligence requirement 优先情报需求

PL phase line 调整线

PR personnel recovery 人员营救

PZ pickup zone 接载区

QC quality control 质量控制

ROE rules of engagement 交战规则

ROZ restricted operations zone 限制飞行区

RW rotary-wing 旋转翼

SAM surface to air missile 地对空导弹

SCAR strike coordination and reconnaissance 打击协调与侦察

SEAD suppression of enemy air defense 压制敌方防空

SOP standard operating procedure 标准操作流程

SPO support operations officer 支援作战军官

SSB security and support battalion 警戒与支援营

SWT scout weapons team 侦察武器小组

TAB-A theater aviation brigade（assault）战区航空（陆航）旅（突击）

TAB-GS theater aviation brigade（general support）战区航空（陆航）旅（全般支援）

TACON tactical control 战术控制

TAIS tactical airspace integration system 战术空域集成系统

TAOG theater airfield operations group 战区机场作业大队

TASMG theater aviation sustainment maintenance group 战区航空（陆航）保障维修大队

TCF tactical combat force 战术战斗部队

TRP target reference point 目标参考点

TTP tactics，techniques，and procedures 战术、技术和程序

UAS unmanned aircraft system 无人机系统

USAR United States Army Reserve 美国陆军预备役

Terms

Air assault 空中突击

The movement of friendly assault forces by rotary-wing or tiltrotor aircraft to engage and destroy enemy forces or to seize and hold key terrain.

通过旋转翼或倾转旋转翼飞机运送己方突击部队与敌交战并摧毁敌方部队、控制关键地形。

Air-ground operations 空地作战

The simultaneous or synchronized employment of ground forces with aviation maneuver and fires to seize, retain, and exploit the initiative.

地面部队与航空机动和火力的同步或协同运用，以夺取、保持并利用主动权。

Air movement 空中运输

Air transport of units, personnel, supplies, and equipment including airdrops and air landings.

空运部队、人员、补给和装备，包括空投和空降。

Area reconnaissance 地域侦察

A form of reconnaissance that focuses on obtaining detailed information about the terrain or enemy activity within a prescribed area.

一种侦察形式，重点是获取指定地域内有关地形和敌人活动的详细信息。

Army personnel recovery 陆军人员营救

The military efforts taken to prepare for and execute the recovery and reintegration of isolated personnel.

准备并实施营救失散人员，使其回归部队所采取的军事行动。

Consolidation area 巩固地域

The portion of the commander's area of operations that is designated to facilitate the security and stability tasks necessary for freedom of action in the close area and to support the continuous consolidation of gains.

指挥官作战区域中的某一部分，旨在帮助执行必要的警戒和维稳任务，以确保近距离地域的行动自由，并支持持续巩固已取得的战果。

Coordinating altitude 协调高程

An airspace coordinating measure that uses altitude to separate users and as the transition between different airspace control elements.

一种用高程区分空域使用者的空域协调措施，也是不同空域管制部门之间的交接高程。

Coordination level 协调高度

A procedural method to separate fixed- and rotary-wing aircraft by determining an altitude below which fixed-wing aircraft normally will not fly.

一种程序方法,通过确定一个固定翼飞机通常不会低于其飞行的高度,以区分固定翼飞机和旋翼飞机。

Cover 掩护

A security task to protect the main body by fighting to gain time while also observing and reporting information and preventing enemy ground observation of and direct fire against the main body.

警戒任务之一,通过战斗争取时间并观察报告信息,同时防止敌军地面观察和直接火力打击,以保护主力部队。

Decisive operation 决定性作战

The operation that directly accomplishes the mission.

直接完成任务的作战。

Forward arming and refueling point 前沿装弹和加油点

A temporary facility, organized, equipped, and deployed to provide fuel and ammunition necessary for the employment of aviation maneuver units in combat.

临时组建、装备和部署的设施,为战斗中的航空机动部队运用提供必要的燃油和弹药。

Guard 警卫

A security task to protect the main force by fighting to gain time while also observing and reporting information and preventing enemy ground observation of and direct fire against the main body. Units conducting a guard mission cannot operate independently because they rely upon fires and functional and multifunctional support assets of the main body.

警戒任务之一,通过战斗争取时间并观察报告信息,同时防止敌军地面观察和直接火力打击,以保护主力部队。执行警卫任务的部队不能独立作战,因为他们依赖主力部队提供火力以及职能性和多功能支援资产。

Hybrid threat 混合威胁

The diverse and dynamic combination of regular forces, irregular forces, terrorist forces, or criminal elements unified to achieve mutually benefitting threat effects.

正规军队、非正规军队、恐怖分子团伙、罪犯团伙的多样化、动态化组合联合,以实现互相促进的威胁效果。

Information collection 情报收集

An activity that synchronizes and integrates the planning and employment of sensors and assets as well as the processing, exploitation, and dissemination systems in direct support of current and future operations.

一种同步并集成传感器及资产规划与使用，以及信息处理、开发和分发系统运作的活动，以直接支援当前和未来作战。

Information environment 信息环境

The aggregate of individuals, organizations, and systems that collect, process, disseminate, or act on information.

收集、处理、分发信息或依据信息采取行动的个人、组织和系统的总和。

Kill box 杀伤区

A three-dimensional permissive fire support coordination measure with an associated airspace coordinating measure used to facilitate the integration of fires.

一种三维许可火力支援协调措施，附带相关的空域协调措施，用于促进火力集成。

Main command post 基本指挥所

A facility containing the majority of the staff designed to control current operations, conduct detailed analysis, and plan future operations.

一种包括大多数参谋人员的设施，旨在控制当前作战，实施详细分析和计划未来作战。

Main effort 主力

A designated subordinate unit whose mission at a given point in time is most critical to overall mission success.

一支指定的下属部队，其在某一特定时间点的任务对整体任务成功最为关键。

Manned unmanned teaming 有人-无人编队

The integrated maneuver of Army Aviation RW and UAS to conduct movement to contact, attack, reconnaissance, and security tasks.

陆军航空兵旋转翼飞机和无人机的综合机动，用于执行接敌运动、攻击、侦察和警戒任务。

Operational environment 作战环境

A composite of the conditions, circumstances, and influences that affect the employment of capabilities and bear on the decisions of the commander.

影响指挥官能力运用的各种条件、情况和因素的综合体，其对指挥官的决策有重要影响。

Operational reach 作战范围

The distance and duration across which a joint force can successfully employ military capabilities.

联合部队能够成功运用军事能力的距离和持续时间。

Reconnaissance 侦察

A mission undertaken to obtain, by visual observation or other detection methods，information about the activities and resources of an enemy or adversary，or to secure data concerning the meteorological，hydrographic or geographic characteristics of a particular area.

一项通过视觉观察或其他探测方法，获取关于敌方或对手的活动和资源的信息，或者收集特定区域的气象、水文或地理特征数据的任务。

Reconnaissance in force 武装侦察

A deliberate combat operation designed to discover or test the enemy's strength，dispositions，and reactions or to obtain other information.

一种有计划的战斗行动，旨在发现或测试敌方的实力、部署和反应，或者获取其他信息。

Route reconnaissance 路线侦察

A directed effort to obtain detailed information of a specified route and all terrain from which the enemy could influence movement along that route.

一种指令行动，旨在获取指定路线及其周边所有地形的详细信息，这些地形可能被敌方用以影响沿该路线的行动。

Screen 屏护

A security task that primarily provides early warning to the protected force.

主要为受保护部队提供预警的警戒任务。

Security tasks 警戒任务

Those tasks performed by commanders to provide early and accurate warning of enemy operations，to provide the forces being protected with time and maneuver space within which to react to the enemy，and to develop the situation to allow commanders to effectively use their protected forces.

由指挥官执行的任务，旨在提供敌方行动的早期和准确预警，为受保护的部队提供反应时间和机动空间，并发展态势以使指挥官能够有效利用受保护的部队。

Shaping operation 塑造行动

An operation that establishes conditions for the decisive operation through effects on the enemy，other actors，and the terrain.

一种通过影响敌方、其他行为体以及地形，为决定性行动创造条件的作战行动。

Space domain 太空域

The area above the altitude where atmospheric effects on airborne objects become negligible.

某高度之上区域，此处大气效应对空中物体的影响可以忽略不计。

Special reconnaissance 特种侦察

Reconnaissance and surveillance actions conducted as a special operation in hostile, denied, or diplomatically and/or politically sensitive environments to collect or verify information of strategic or operational significance, employing military capabilities not normally found in conventional forces.

在敌对、拒止或外交和/或政治敏感的环境中,进行侦察和监视等特种作战行动,以收集或验证具有战略或作战意义的信息,常规部队通常不具有执行此类行动的军事能力。

Strike coordination and reconnaissance 打击协调与侦察

A mission flown for the purpose of detecting targets and coordinating or performing attack or reconnaissance on those targets.

一种飞行任务,目的是探测目标并协调或对这些目标进行攻击或侦察。

Support area 支援地域

In contiguous areas of operations, an area for any commander that extends from its rear boundary forward to the rear boundary of the next lower level of command.

在连续的作战区域内,任何指挥官的责任区域从其后界向前延伸到下一级指挥单位的后界。

Supporting effort 支援部队

A designated subordinate unit with a mission that supports the success of the main effort.

指定的下级部队,其任务是支援主力取得胜利。

Sustaining operation 持续性作战行动

Those operations at any echelon that enable the decisive operation or shaping operations by generating and maintaining combat power.

在任何层级上,通过生成和维持战斗力支持决定性行动或塑造行动的作战行动。

Tactical command post 战术指挥所

A facility containing a tailored portion of a unit headquarters designed to control portions of an operation for a limited time.

包括部队指挥部某一特定部分的设施,目的是在有限时间内控制部分作战行动。

Threat 威胁

Any combination of actors, entities, or forces that have the capability and intent to harm the forces, national interests, or the homeland.

任何有能力和意图对部队、国家利益或国土造成伤害的行为体、实体或力量的组合。

Zone reconnaissance 地带侦察

A form of reconnaissance that involves a directed effort to obtain detailed information on all routes，obstacles，terrain，and enemy forces within a zone defined by boundaries.

一种侦察形式，涉及有目的地获取某一边界限定地带内的所有路线、障碍、地形和敌军的详细信息。

References

1. 张德和等著,唐海民,徐章俊译. The PLA Army Aviation Corps. 北京：五洲传播出版社,2013.

2. 张德和,蒲先斌,张晓东等. 中国军队系列：中国人民解放军陆军航空兵[M]. 北京：五洲传播出版社,2013.

3. 王传经,(英)Simon Mellor-Clark,(英)Yvonne Baker de Altamirano. 军事英语听说教程. 北京：外语教学与研究出版社,2023.

4. 160th Special Operations Aviation Regiment (Night Stalkers-160th SOAR). (2018). Retrieved January 5, 2024, from https://militaryleak.com/2018/01/02/160th-special-operations-aviation-regiment-night-stalkers-160th-soar

5. Boyne, W. J. (n. d.). Helicopter：Fact, History, and Types. Retrieved October 7, 2024, from https://www.britannica.com/technology/helicopter

6. China Military Online. (2019). Deeply Impressed by Chinese Peacekeepers in Africa. Retrieved October 27, 2024, from http://eng. chinamil. com. cn/CHINA_209163/Features_209191/9707042. html

7. Dore, T. (2021). History of the 128th Aviation Brigade. Army Aviation, 2021(1), 26 - 27.

8. Federal Aviation Administration. (n. d.). Radio Communications Phraseology and Techniques. In Aeronautical Information Manual (AIM) [Web page]. Retrieved October 13, 2024, from https://www. airresearch. com/Pilots/AIM-08/Chap4/aim0402. html

9. Headquarters, Department of the Army. Field Manual No. 1 - 100 Army Aviation Operations. 1997.

10. Headquarters, Department of the Army. Field Manual No. 3 - 04 Army Aviation. 2020.

11. Husseini, T. (2019). Advanced Military Helicopters：How Function Dictates Capability. Retrieved April 17, 2022, from https://www. airforce-technology. com/features/advanced-military-helicopters/?cf-view

12. King, J. C., III, Black, P. A., & Burney, N. D. (2020). Warrant Officer and Enlisted Maintenance Training. Army Aviation, 2020(1), 24.

13. Lange, K. (2020). Why Army Helicopters Have Native American Names. Army Flier, 70(5), 7.

14. NPR. (2013). What a Downed Black Hawk in Somalia Taught America [Web page]. Retrieved October 27, 2024, from https://www.npr.org/2013/10/05/229561805/what-a-downed-black-hawk-in-somalia-taught-america

15. Paszak, P. (2020). Are Military Drones the Future of the Chinese Army? Warsaw Institute. Retrieved January 31, 2022, from https://warsawinstitute.org/military-drones-future-chinese-army

16. UN Peacekeeping. (n.d.) Somalia-UNOSOM II Background. Retrieved October 27, 2024, from https://peacekeeping.un.org/sites/default/files/past/unosom2backgr2.html

17. Wiesner, I. (2017). A Sociology of the Drone. Journal of Military and Strategic Studies, 18(1), 42 – 59.

18. Wood, P. (2018). PLA Army Aviation Brigades' Training Now Includes Urban Operations. Retrieved October 13, 2024, from https://community.apan.org/wg/tradoc-g2/fmso/m/oe-watch-articles-singular-format/266816

图书在版编目(CIP)数据

航空英语综合教程 / 王欣然，贡卫东主编. -- 南京 ：
南京大学出版社，2024.8. -- ISBN 978 - 7 - 305 - 28162 - 4

Ⅰ. V2

中国国家版本馆 CIP 数据核字第 2024EN7986 号

出版发行　南京大学出版社

社　　　址　南京市汉口路 22 号　　　　　　邮　编　210093

书　　　名　**航空英语综合教程**
　　　　　　HANGKONG YINGYU ZONGHE JIAOCHENG

主　　　编　王欣然　贡卫东

责任编辑　刁晓静　　　　　　　　　编辑热线　025 - 83592123

照　　　排　南京南琳图文制作有限公司

印　　　刷　丹阳兴华印务有限公司

开　　　本　787 mm×1092 mm　1/16　印张 13.5　字数 280 千

版　　　次　2024 年 8 月第 1 版　2024 年 8 月第 1 次印刷

ISBN 978 - 7 - 305 - 28162 - 4

定　　　价　46.00 元

网址：http://www.njupco.com

官方微博：http://weibo.com/njupco

官方微信号：njupress

销售咨询热线：(025) 83594756